T0313347

Artificial Intelligence
in Wireless Robotics

RIVER PUBLISHERS SERIES IN INFORMATION SCIENCE AND TECHNOLOGY

Series Editors:

K. C. CHEN
National Taiwan University, Taipei, Taiwan
and
University of South Florida, USA

SANDEEP SHUKLA
Virginia Tech, USA
and
Indian Institute of Technology Kanpur, India

Indexing: All books published in this series are submitted to the Web of Science Book Citation Index (BkCI), to SCOPUS, to CrossRef and to Google Scholar for evaluation and indexing.

The "River Publishers Series in Information Science and Technology" covers research which ushers the 21st Century into an Internet and multimedia era. Multimedia means the theory and application of filtering, coding, estimating, analyzing, detecting and recognizing, synthesizing, classifying, recording, and reproducing signals by digital and/or analog devices or techniques, while the scope of "signal" includes audio, video, speech, image, musical, multimedia, data/content, geophysical, sonar/radar, bio/medical, sensation, etc. Networking suggests transportation of such multimedia contents among nodes in communication and/or computer networks, to facilitate the ultimate Internet.

Theory, technologies, protocols and standards, applications/services, practice and implementation of wired/wireless networking are all within the scope of this series. Based on network and communication science, we further extend the scope for 21st Century life through the knowledge in robotics, machine learning, embedded systems, cognitive science, pattern recognition, quantum/biological/molecular computation and information processing, biology, ecology, social science and economics, user behaviors and interface, and applications to health and society advance.

Books published in the series include research monographs, edited volumes, handbooks and textbooks. The books provide professionals, researchers, educators, and advanced students in the field with an invaluable insight into the latest research and developments.

Topics covered in the series include, but are by no means restricted to the following:

- Communication/Computer Networking Technologies and Applications
- Queuing Theory
- Optimization
- Operation Research
- Stochastic Processes
- Information Theory
- Multimedia/Speech/Video Processing
- Computation and Information Processing
- Machine Intelligence
- Cognitive Science and Brain Science
- Embedded Systems
- Computer Architectures
- Reconfigurable Computing
- Cyber Security

For a list of other books in this series, visit www.riverpublishers.com

Artificial Intelligence in Wireless Robotics

Kwang-Cheng Chen

University of South Florida, USA

Routledge
Taylor & Francis Group

LONDON AND NEW YORK

Published 2020 by River Publishers
River Publishers
Alsbjergvej 10, 9260 Gistrup, Denmark
www.riverpublishers.com

Distributed exclusively by Routledge
4 Park Square, Milton Park, Abingdon, Oxon OX14 4RN
605 Third Avenue, New York, NY 10017, USA

Artificial Intelligence in Wireless Robotics / by Kwang-Cheng Chen.

Routledge is an imprint of the Taylor & Francis Group, an informa business

ISBN 978-87-7022-118-4 (print)

While every effort is made to provide dependable information, the publisher, authors, and editors cannot be held responsible for any errors or omissions.

Contents

Preface

Robotics has been developed for decades. Initially, robotics was pretty much treated by control engineering and mechanical engineering. Then, computer engineering was amended, and later more computer science, particularly artificial intelligence (AI), was incorporated. This book intends to include more technological components from wireless communication for sensors and multi-robot systems, to form a new technological front of the *wireless robotics*. Such new technological components enrich the AI in wireless robotics to result in the vision of this book.

The manuscript of this book was developed based on the class note of a new graduate course *Robotics and AI* offered by the author at the University of South Florida, which is well suitable for the first-year graduate students and senior undergraduate students only with prior knowledge in undergraduate probability and matrix algebra, in addition to basic programming. The new aspect of this book is to introduce the role of wireless communication technology enhancing the AI that is used for robotics. Consequently, the title of this book is *Artificial Intelligence in Wireless Robotics*. There are many application scenarios of robotics and this book primarily focuses on autonomous mobile robots and robots requiring wireless infrastructure such as those in a (networked) smart factory. It is also noted that robotics involves multi-disciplinary knowledge, mostly in electrical engineering, computer science, computer engineering, and mechanical engineering. Considering a decent number of pages, this book is not going to cover every aspect of robotics. Instead, this book is prepared for readers and students without any prior knowledge in robotics, by introducing AI and wireless factors into robotics.

The book consists of 10 chapters. The first chapter presents basic knowledge in robotics and AI. Chapters 2 and 3 provides basic knowledge in AI search algorithms and machine learning techniques. Chapter 4 first briefs the statistical decisions and then Markov decision processes. Chapter 5 plays a central role to introduce reinforcement learning. Chapter 2 to 5 appears more "computer science" for AI. Chapter 6 supplies the fundamental knowledge in estimation, which is useful to establish the belief of a robot, and to develop more techniques (often with wireless) to enrich the AI in robots. Chapter 7 further applies the knowledge of estimation to a critical problem in autonomous mobile robots (AMRs), localization, which is also related to the robot pose problem. Chapter 8 presents the planning for a robot to elevate the intelligence level of robots. Chapter 9 first orients the vision for robots, particularly AMRs, then considers fuse the information from multiple kinds of sensors as multi-modal fusion. Chapters 6, 7, and 9, may be viewed as the *signal processing* approach to enhance the AI in robotics, a more "electrical engineering" view. Chapter 10 briefly introduce the multi-robot systems. Instead of popular study on a swarm of tiny robots, we focus more on the collaborative robots while each of them equips with good computing capability. It also suggests the potentially important role of wireless communication in robotics, which is briefly introduced. Under a decent number of pages, *AI in wireless robotics* is oriented from every aspect, with enrichment from wireless and signal processing technology.

At the end of each chapter, some further reading for more depth and details would be suggested. The exercises are labelled by ▶ and the computer exercises are labelled by ■ in front. These exercises are the integral part of the text toward deeper understanding. The computer exercises usually take non-trivial efforts but there are a lot of funs according to the feedback from students in the class. They also significantly help your in-depth understanding of the technical contents. Please enjoy them.

Any project of serious efforts relies on a lot of support behind. The author would like to thank two consecutive Department Chairs, Tom Weller and Chris Ferekides, for their encouragement and support to offer a new graduate course on this subject, such that the author can transform the class note to the book manuscript. During the preparation of the book manuscript, I would like to appreciate the assistance from my graduate students Eisaku Ko and Hsuan-Man Hung at the National Taiwan University; Ismail Uluturk, Zixiang Nie, post-doc Amanda Chiang, and Teaching Assistant Zhengping Luo at the University of South Florida; Pengtao Zhao and Yingze Wang at the Beijing Univerity of Post and Telecommunications; undergraduate students Jose Elidio Campeiz (University of South Florida) and Daniel T. Chen (Case Western Reserve University) for their proof reading. Of course, graduate students from different departments in the College of Engineering, University of South Florida, who took the graduate course *Robotics and AI*, supply tremendous valuable feedbacks and comments, which inevitably improve this book. Professor Qimei Cui arranged a summer course allowing me to teach the manuscript of this book to more than 70 students with more students sitting in, at the Beijing University of Post and Telecommunications in 2019, which helps a lot to get more feedback. Of course, Rajeev and Junko from the River Publisher assisted a lot in the final preparation of this book. Finally, without the plausible caring from my wife Kristine, it is not possible for me to focus on writing.

K.-C. Chen, Lutz, Florida

List of Figures

List of Tables

List of Abbreviations

AGC	Automatic gain control
AI	Artificial Intelligence
AMR	Autonomous mobile robot
ANN	Artificial neural networks
AOA	Angle-of-arrival
AR	Auto-regressive
ASK	Amplitude shift keying
AV	Automatic Vehicle
AWGN	Additive white Gaussian noise
BER	Bit error rate
BFS	Breadth-first search
BLUE	Best linear unbiased estimator
BPSK	Binary frequency shifted keying
BPSK	Binary phase shift keying
BS	Base station
CART	Classification and regression tree
CDF	Cumulative distribution function
CEP	Conditional exhaustive planning
CPU	Central Processing Unit
CSMA	Carrier sense multiple access
CSP	Constraint Satisfaction Problem
DAG	Directed acyclic graph
DFS	Depth-first search
DL	Deep learning
DLC	Data link control
DNN	Deep Neural Networks
DoF	Degrees of freedom
EKF	Extended Kalman filter
EKF-SLAM	Extended Kalman filter SLAM
EM	Expectation-maximization
ERM	Empirical risk minimization

FIFO	First-in-first-out
FL	Federated learning
FSK	Fequency shift keying
FSM	Finite-state machine
FSMC	Finite-state Markov chain
GLRT	Generalized likelihood ratio test
GMM	Gaussian mixture model
GPS	Global Positioning System
HFL	Horizontal federated learning
HMM	Hidden Markov model
I2V	Infrastructure-to-vehicle
IA	Instantaneous assignment
ICA	Independent component analysis
IDA*	Iterative-deepening A*
KNN	K nearest neighbors
LASSO	Least absolute shrinkage and selection operator
LBT	Listen-before-transmission
LEO	Localization error outage
LIFO	Last-in-first-out
Linear MMSE	Linear minimum mean squared error estimator
LLC	Logic link control
LOS	Line of sight
LP	Linear programming
LS	Least-squares
LSE	Least squared error
LTI	Linear time-invariant
MA	Moving average
MA*	Memory-bounded A*
MAB	Multi-armed bandit
MAC	Medium access control
MANET	Mobile ad hoc network
MAP	Maximum a posteriori
MAS	Multi-agent system
MC methods	Monte Carlo methods
MDP	Markov decision process
ML	Machine learning
MLE	Maximum likelihood estimation
MMAE	Minimum mean absolute error estimate
MMSE	Minimum mean squared error estimator

MR	Multi-robot
MRS	Multi-robot systems
MRTA	Multi-robot task allocation
MSE	Mean squared error
MT	Multi-task
mTSP	Multiple traveling salesmen problem
MVUE	Minimum variance unbiased estimator
NetMAS	Networked MAS
NLOS	Non-line of signt
NN	Neural Network
OA	Optimal assignment
OFDM	Orthogonal Frequency-division Multiplexing
OSI	Open system interconnection
PCA	Principal component analysis
PMF	Probability mass function
POMDP	Partially observed MDP
PSK	Phase shift keying
QAM	Quadrature amplitude modulation
QPSK	Quadrature phase shift keying
RADAR	Radio detection and ranging
RBFS	Recursive best-first search
RF	Radio frequency
RL-DT	Reinforcement Learning with Decision Trees
RNN	Recurrent neural network
R-nodes	Reducing nodes
ROC	Receiver operating curve
RRU	Radio Resource Unit
RSS	Received signal strength
RSS	Residual sum of squares
rt-ALOHA	Real-time ALOHA
SARSA	State-action-reward-state-action
SLAM	Simultaneous localization and mapping
SMA*	Simplified MA*
S-nodes	Sensing nodes
SNR	Signal-to-noise ratio
SR	Single-robot
SRM	Structural risk minimization
SSR	Sum of squares regression
SST	Sum of squares total

ST	Single-task
SVMs	Support vector machines
TA	Time-extended assignment
TD learning	Temporal-difference learning
TDE	Time delay estimation
TDMA	Time Division Multiple Access
TDOA	Time-difference-of-arrivals
TOA	Time-of-arrival
TSP	Traveling Salesman problem
UAV	Unmanned aerial vehicle
UCB	Upper confidence bounds
UE	User entity
uRLLC	Ultra reliable low-latency communication
V2I	Vehicle-to-infrastructure
V2I2V	Vehicle-to-infrastructure-to-vehicle
V2V	Vehicle-to-vehicle
VFL	Vertical federated learning
WLAN	Wireless local area network
WSNs	Wireless sensor networks

1

Introduction to Artificial Intelligence and Robotics

Advances in Artificial Intelligence (AI) technology and related fields have opened up new markets and new opportunities for progress in critical areas such as health, education, energy, economic inclusion, social welfare, and the environment. In recent years, machines have surpassed humans in the performance of certain tasks related to intelligence, such as aspects of image recognition. Experts forecast that rapid progress in the field of specialized artificial intelligence will continue. Although it is unlikely that machines will exhibit broadly-applicable intelligence comparable to or exceeding that of humans in the next 20 years, it is to be expected that machines will continue to reach and exceed human performance on more and more tasks.

AI-driven automation will continue to create wealth and expand the American economy in the coming years, but, while many will benefit, that growth will not be costless and will be accompanied by changes in the skills that workers need to succeed in the economy, and structural changes in the economy.......

Executive Office of the President, United States of America, December 20, 2016.

1.1 Common Sense Knowledge of AI, Cybernetics, and Robotics

Artificial intelligence (AI) is generally known as the synthesis and analysis of computational agents that act intelligently. An agent is a machine or any computational entity that can act. The scientific goal of AI is to understand the

1

principles facilitating intelligent behavior in natural or man-made systems, with primary functionalities:

- analysis of natural and artificial agents
- formulating or testing hypotheses to construct intelligent agents
- designing, establishing, experimenting with computational systems that perform tasks of intelligence

Example: A digital communication receiver is an intelligent agent to determine from the hypotheses (say, H_0 and H_1) of the transmitted digital signals by processing the received waveforms. There is no surprise, since 1980's, engineers implement communication receivers using (computer) processors and/or digital signal processors that are special-purpose processors of even higher processing power in million instructions per second (MIPS) or floating-point operations per second (FLOPS).

Example: Computer agents of touch panels as the interface with humans are common in many situations such as airline check-in in Figure 1.1 and ATM machines in banking industry.

Example: Robots to relieve the need of human working force are considered highly potential applications, say smart manufacturing. Figure 1.2 shows robots to pick up strawberry.

Prior to the Von Neumann architecture in 1945 toward the birth of human's first stores-program computer, the concept of AI had been brought

Figure 1.1   Computer agents responsible for airline customers' check-in.

Figure 1.2 Robots to pick up strawberries being considered to save the agriculture; http://www.wbur.org/npr/592857197/robots-are-trying-to-pick-strawberries-so-far-theyre-not-very-good-at-it.

up, ironically thanks to the need of war technology. During the World War II, the roots of AI had been was fostered in different ways but aimed at supplying technological edge to win the war. Among them, the most well known pioneers could be Alan Turing (1912–1954) and Norbert Wiener (1894–1964).

A. Turing created the famous *Turing machine* as a model of computation, which is believed to decode the sophisticated *Enigma* crypto-graphical machine. Furthermore, the Turing test is widely accepted to test whether a computational machine is artificially intelligent.

Example: The Turing test consists of an imitation game where an interrogator can ask a witness any question. If the interrogator can not distinguish the witness from a human, the witness must be intelligent. Due to the man-machine interface, the original test was considered via a text interface. Today, if you want to download a paper from digital library or a band statement from email notice, image test following the principle of Turing test will be employed to make sure that you are not a robot trying in an exhaustive manner. Such a test illustrated in Figure 1.3 serves a fundamental role in modern cybersecurity.

In addition to the Wiener filter that is the optimal filtering of stationary random signals, N. Wiener has been considered to create the *cybernetics*,

Figure 1.3   (left) "I'm not a robot" test example of letters (right) google's reCAPTCHA using images to test.

which deals with control and communication in the animal and in the machine according to his original definition in 1948. State-of-the-art cybernetics can be viewed as automation and control of any system that is heavily related to robotics. Back to World War II, Germans' ballistic missiles V-1 and more mature V-2 could be considered as the first kind of military robots in successful operation. Automatic control emerged as the center of cybernetics. Quickly after the invention of electronic computers, computing has joined control and communication into cybernetics, while automation and man-machine interaction remains highly interesting in technology developmment.

According to N. Wiener, the first industrial revolution was the devaluation of human arms by the competition of machinery. The second industrial revolution was to devalue the human brain, at least in its simpler and more route decisions. A good example might be the assembly line in manufacturing. It is widely believed that the introduction of computers, particularly personal computers and thus Internet, invokes the third industrial revolution to replace a lot of straightforward-expertise tasks, say spreadsheet like Excel to greatly simplify calculations and thus accounting. When AI is brought into technological solutions, many people believe the fourth industrial revolution (i.e. Industry 4.0) happening, which will be lightly touched in Chapter 10.

John McCarthy, PhD Princeton University, after moving to Stanford, settled at the Dartmouth College. McCarthy convinced Minsky, Claude Shannon, and Nathaniel Rochester to bring together U.S. researchers interested in automata theory, neural networks, and the study of intelligence, for a two-month workshop at Dartmouth in the summer of 1956. It is the official birth of AI. After years' evolution, there are four categories of AI indicated in the book of *Artificial Intelligence: A Modern Approach* by S. Russell and P. Norvig:

- Thinking Humanly
- Acting Humanly
- Thinking Rationally
- Acting Rationally

Within the application scenarios of robotics, industry 4.0 and smart manufacturing, and autonomous systems/machines, we primarily focus on "Acting Rationally" in this book and semantic computing and intelligence is not included in general.

To fully comprehend knowledge of Robotics and AI, there involve three branches of mathematics:

Logic: The word and idea of *algorithm* came from a Persian mathematician in the 9th century. Starting by Boole, efforts were under way to formalize general mathematical reasoning as logical deduction by late 19th century.

Computation: The knowledge of *incompleteness theorem* implies that some functions cannot be represented by an algorithm, that is, not computable, which motivates A. Turing trying to characterize exactly which functions are *computable* (i.e. capable of being computed). It is stated that Turing machine is capable of computing any computable function. Then, *tractability* arises for greater impact. If the time required to solve instances of the problem grows exponentially with the size of the instances, such a problem is known as intractable. The theory of *NP-completeness* provides a method to recognize an intractable problem.

Probability and Statistics: AI, particularly robotics, has to deal with uncertainty in cognition and decision, which probabilistic modeling becomes useful. Hypothesis testing and estimation, major categories of problems in statistics, establishes the foundation of decision, statistical perception, communication, control, and computing for reasoning.

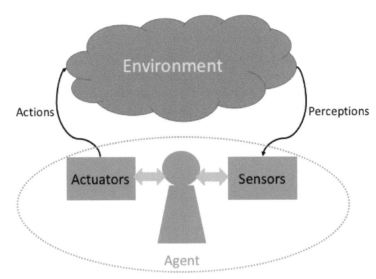

Figure 1.4   An Agent Interacts with the Environment.

1.2 Intelligent Agents

Among wide scope of AI, AI on the agent is of most interest and related to robotics. An agent acts in an environment, and has capability of perception, reasoning, and acting. If the agent is a robot, perception is typically facilitated by some sensors such as cameras or radars; acting relies on the actuators; reasoning is usually realized by a computational engine executing machine learning and decisions. Figure 1.4 illustrates such a scenario for a robot. The rest of the book focuses on the knowledge to design such an intelligent agent and their interactions as a *multi-agent system*.

An agent together with the environment is called a *world*. In addition to the physical robots, an agent of an advice-giving computing capability with a human providing perception and carrying out the task is an *expert system*. An agent who has a program acting purely computational environment is a *software agent*. Figure 1.4 applies too.

1.2.1 The Concept of Rationality

The agent in our scope is to act rationally, which means that a *rational agent* should do right things. Based on the agent's perceptions on the state of environment (which might not be the true state of the environment), a rational agent makes a right decision or takes a right action responding to the environment. Based on the action determined by the agent, actuators interact on the environment, which please note might not be precise in practical engineering. A core problem of the agent's AI is how to take a right action for this rational agent.

It is known from *statistical decision theory* that a rational decision should consider the *consequences* of the decision and the *priori probability* associated with the decision process. The consequences could be represented by a *performance measure*. The performance measure of consequence can be named in different ways, but usually preferred the existence of mapping to real values for the sake of computation. Due to the uncertainty, by introducing probabilistic or statistical means, rational agents usually maximize the *expected performance*.

Game Theory: In the pioneer exploration by Von Neumann and Morganstern, the utility has been introduced to as the performance measure to make decisions. It is modeled via the *utility function*, u, which assigns a utility value $u(x) \in \mathbb{R}$ for each consequence $x \in \mathfrak{X}$. To model the uncertainty of state, we assume a probability distribution p on the state space $\mathfrak{S}$, which is known as the *decision under risk* by Von Neumann and Morganstern. This distribution can be obtained by objective statistics or subjective cognition. Denoting the decision a based on the observation x, the *expected utility* is therefore

$$\mathbb{EU} = \sum_{s \in \mathfrak{S}} p(s) u\left[a(s)\right] \tag{1.1}$$

Statistical Estimation: In parameter estimation, let $\hat{\theta}$ be the estimate of θ, *loss function* L can be defined as follows and then take expectation $\mathbb{E}\left[L\right]$.

- quadratic loss: $L(\theta, \hat{\theta}) = (\theta - \hat{\theta})^2$
- absolute-value loss: $L(\theta, \hat{\theta}) = |\theta - \hat{\theta}|$
- truncated quadratic loss: $L(\theta, \hat{\theta}) = \min\{(\theta - \hat{\theta})^2, d^2\}$

Similarly, in statistical decisions, we may sometimes use the risk function instead of loss function.

Digital Communications: To design a binary digital communication receiver to detect signals "1" or "0", hypothesis testing in statistics will be employed. The *cost* for the detection as "1" when "0" is transmitted is usually treated as 1, counting as one error. Similarly, the cost for the detection as "0" when "1" is transmitted can be treated as 1, again counting as one error. In this sense, the performance measure means the *bit error rate* (BER) of a binary digital communication system.

Please note that it is better to design performance measures according to what one actually wants responding to the environment, rather than how one wishes the agent should behave in the environment.

1.2.2 System Dynamics

Any AI system is generally dynamic in time, particularly in robotics. According to classical mechanics, the motion of a physical object can be represented with six *degrees of freedom*: (a) position in three-dimension space, x, y, z (b) the angles of rotation around x-axis, y-axis, z-axis, that is, $\theta_x, \theta_y, \theta_z$. Newtonian mechanics govern the analytical description of such system dynamics.

The model of the system can be represented by a real-valued function

$$S : X \to Y \tag{1.2}$$

which can be depicted by Figure 1.5 where S denotes a system and $x, y \in \mathbb{R}$. Such a system box with the inputs as functions and the outputs as functions is called an *actor*.

Intuitively, a system is *causal* if its output depends only on the current and past inputs. A system is linear if the superposition property. That is, $\forall x_1, x_2 \in X$ and $\forall a, b \in \mathbb{R}$,

$$S(ax_1 + bx_2) = aS(x_1) + bS(x_2) \tag{1.3}$$

If S is linear and invariant with time, this is a *linear time-invariant* (LTI) system. A system is called *stable* for bounded-input and bounded-output. Figure 1.6 shows a feedback system, where the *error signal* (that is, the difference between input and feedback) actually feeds into the S. Such a system configuration is widely used in control engineering to stabilize the system.

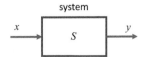

Figure 1.5   Actor model of a system.

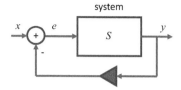

Figure 1.6 Feedback system.

Throughout of this book or system engineering, the concept of *state* has been adopted. The state of a system is intuitively defined as its condition at a particular point in time. The state generally affects the manner that a system reacts to inputs. A *state machine* forms a system model with discrete dynamics mapping reactions from inputs to outputs. In case the the set of states is finite, the state machine is a *finite-state machine* (FSM). The *state transition diagram* based on the state-machine can well represent the system dynamics. The states are usually depicted as rectangles or circles, and conditions/actions are usually depicted as directed arcs. The following example depicts an example for the state transition diagram of FSM.

Example: A cognitive radio is considered as a radio terminal with the capability of intelligence, which senses the spectrum availability before transmission. The default state is in the "reception" mode. If there is a packet to transmit, the radio terminal first senses the spectrum opportunity (i.e. availability of spectrum), into the "sensing" mode. If the spectrum opportunity exists, the radio turns to the "transmission" mode. Once transmission is completed, the system falls back to the "reception" mode. If no spectrum opportunity, it stays in the "sensing" mode. If the transmission fails and the terminal knows, it falls back to the "sensing" mode. Figure 1.7 summarizes the state transition diagram of cognitive radio FSM.

Remark: The concept of FSM is useful in computing, data flows of computation, logic operation or circuitry, software flow, control, and communications or networking. It will be applied in many later chapters.

1.2.3 Task Environments

For an agent to take a right action, it is related to the sensors to detect the state of environment. A task environment is effectively *fully observable* if the sensors can detect all aspects relevant to the selection of action, where relevance is dependent upon performance measure. The nice aspect of fully observable environment implies no need for an agent to keep internal state

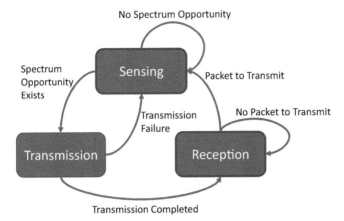

Figure 1.7   State transition diagram for the operation of a cognitive radio.

for the environment in its learning and decision process. However, since the target process inside the environment can be *hidden*, or the sensors are noisy or inaccurate, the environment is thus *partially observable*. In some cases, the environment might be *unobservable*, but it is still possible to achieve agent's goal.

In an episodic task environment, the agent's experience consists of atomic episodes. In each episode the agent receives perception and then performs an action. Crucially, the next episode does not depend on the actions taken in previous episodes. For example, a robot in an assembly line functions in episodic manner. However, in sequential environments, the current decision can affect future decisions. A chess playing agent operates in sequential environment. For both cases, short-term actions can have long-term consequences. A collection of agent's decisions at episodes or instances is called its *policy*.

For an agent to formulate intelligent actions and to effectively compute, the concept of *state space* is introduced. Th information in a *state* allows predictive description useful to action. An appropriate action can be obtained by search for the entire state-space or any computationally effective method achieving similar purpose by assuming:

- The agent has perfect knowledge about the state space, and a plan to observe the state (i.e. full observability).
- the agent knows the consequences of actions.
- There exists performance measurement for an agent to determine whether a state satisfy its goal.

A *solution* is a sequence of actions that allow the agent from its current state to reach a state satisfying its goal.

Example: Suppose a delivery robot taking a package from (room) ENB 118 to destination ENB 245. The current and start (or initial) state is ENB 118, and ENB 245 is the state to complete its mission. A state s_n can be defined as the location in front of a room (numbered n) in ENB building, with the starting state as s_{118} and goal state s_{245}. Action a_1 stands for moving to next room and action a_0 stops for goal state. The evaluation of the delivery mission is the steps reaching the goal.

A state-space problem generally consists of

- a set of states
- the start state (or the initial state)
- a set of actions available to the agent in each state
- a goal state, which can be specified as a Boolean function and is true when the state satisfies the goal
- a criterion to specify the quality of an acceptable solution (e.g. time for a delivery robot to complete its task)

State-space serves an effective approach to model many robotic problems.

▶ **Exercise:** In Figure 1.8 regarding the pole balancing problem, suppose we only consider the scenario as a plane, which means that the platform can only move right and left with possible speed at $0, 1, 2, 3, 4, 5$ m/sec, and the pole of uniform mass can only move clockwise or counter-clockwise. Assume the platform can precisely know the angle of the pole that has uniform density (and thus weight distribution). Please design a reinforcement learning algorithm to balance the pole. For easy calculation, assume $g = 10$ m/sec^2 by gravity and no friction. Please define an appropriate state space for this dynamic system.

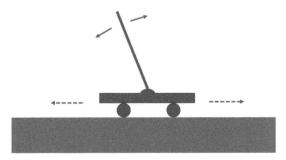

Figure 1.8   Pole balancing.

1.2.4 Robotics and Multi-Agent Systems

Robots are usually considered as physical agents that perform actions to execute tasks by manipulating the physical world. With appropriate physical-cyber interface, robots can be software agents in some cases. Generally speaking, a robot has the following hardware.

- Sensors: Sensors serve as perceptual interface between a robot and the environment. Passive sensors capture true signals that are generated by various sources in the environment, such as cameras and thermometer. Active sensors, on the other hand, send probing signals into the environment, and capture information regarding the environment by analyzing the reflected signals, such as radar, lidar (short for light detection and ranging), sonar, etc. Location sensors based on range sensing are another class of sensors, such as Global Position System (GPS) receivers. The final class of sensors is *proprioceptive sensors*, which supply information to the robot about its motion (e.g. odometry, the measurement of distance traveled) like inertial sensors to complement the functionality of GPS.
- Actuators or Effectors: Actuators supply means for a robot to respond the environment. In particular, effectors allows robots to move or to change their shapes, which can be understood by the concept of *degree of freedom* (DoF). A 6-axis robot arm is illustrated in Figure 1.9.
- On-Board Computing: A robot is equipped with the computing core to act based on the information collected from sensors. Such on-board computing facilitates the functions of artificial intelligence.
- Communication/Networking Device: Current robots usually have limited communication functionality, such as wireless communication to collect data from sensors. Communication and networking among agents is still in the infant stage [1].

Since the true state of the environment might not be directly observed, a primary mission of on-board computing is to recursively update (or estimate) its *belief state* from current belief state and new observation, so as to properly act. Let $\mathbf{X_t}$ be the state (vector) of the environment at time t, $\mathbf{Z_t}$ be the observation (vector) obtained at time t, and A_t be the action taken after obtaining $\mathbf{Z_t}$. The update of belief state can be realized by

$$P(\mathbf{X_t} \mid \mathbf{z}_{1:t+1}, a_{1:t}) = \alpha P(\mathbf{z}_{t+1} \mid \mathbf{X}_{t+1}) \sum_{\mathbf{x}_t}$$
$$P(\mathbf{X}_{t+1} \mid \mathbf{x}_t, a_t) P(\mathbf{x}_t \mid \mathbf{z}_{1:t}, a_{1:t-1}) \quad (1.4)$$

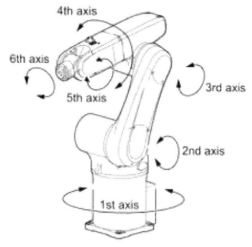

Figure 1.9 6-Axis Robot Arm; https://robotics.stackexchange.com/questions/12213/6-axis-robot-arm-with-non-perpendicular-axes.

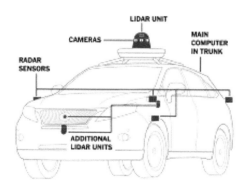

Figure 1.10 Equipment of self-driving car;
Source: New York Times.

which means the posterior over the state variables $\mathbf{X}$ at time $t + 1$ is calculated recursively from the corresponding estimate one time step earlier (i.e. prediction). The probability $P(\mathbf{X_t} \mid \mathbf{z}_{1:t+1}, a_{1:t})$ is know as the *transition model* or *motion model* of the robot, and $P(\mathbf{z}_{t+1} \mid \mathbf{X}_{t+1})$ represents the *sensor model*.

A good example of a robot with intelligence is a self-driving vehicle consisting of various sensors (i.e. lidar, radar, camera), effectors, and power on-board computing, as Figure 1.10.

►**Exercise:** Chapman intends to design a robot dog that can respond to human speech commands, including sit, stand, left, right, go, stop, back. Can you design the hardware architecture and software architecture for such robot dog?

Up to this moment, a single agent has been considered. However, a system may involve multiple agents with interactions, which is known as a *multi-agent system* (MAS). For example, a number of autonomous driving vehicles on the streets, robots in an assembly line in smart factory, or chess-playing agents. The role of these agents can be interactive, cooperative, or competitive. Communication among the agents might be possible, particularly for rational agents. Networked MAS will supply a new frontier to explore in a later chapter of this book.

1.3 Reasoning

The first technical effort toward AI is the reasoning of logic structure, and it can be intuitively facilitated by rule-based operations. Let us proceed in the following. Instead of reasoning explicitly in terms of states, it is common to describe states by *features* that can be defined as *variables* for executions. An *algebraic variable* is a symbol denoting the features in possible worlds. Each algebraic variable X has an associated *domain*, denoted as $dom(X)$.

Example: A Boolean variable has a domain $\{true, false\}$.

Given a set of variables, an *assignment* on the set of variables in a function from the variables to the domains (of variables). A *total assignment* assigns all variables. A possible *world* can be defined on a total assignment.

Example: The variable *ClassTimeMon* denotes the starting time of a class on Monday.
$$dom(ClassTimeMon) = \{8, 9 : 30, 11, 12 : 30, 14, 15 : 30, 17\}$$

Example: There are two variables, $dom(A) = \{0, 1, 2\}$ and $dom(B) = \{t, f\}$. There exist 6 possible worlds.

$$w_0 = \{A = 0, B = t\}$$
$$w_1 = \{A = 1, B = t\}$$
$$w_2 = \{A = 2, B = t\}$$
$$w_3 = \{A = 0, B = f\}$$
$$w_4 = \{A = 1, B = f\}$$
$$w_5 = \{A = 2, B = f\}$$

For many domains, not all possible assignments (of values) to variables are permissible. A *constraint* specifics legal combinations of assignments to some variables. A set of variables can be viewed as a *scope*, S. A *relation* on S is a function from assignments on S to $\{true, false\}$, which specifies whether each assignment is permissible. A *constraint*, c, consists of a scope S and a relation on S. A possible world w satisfies a set of constraints, and the possible world is a *model* of the constraints.

Example: In an intelligent factory, there are robots in one automated assembly line, one robot to drill holes, another robot to screw up connectors, and the final robot to inspect connections. The starting times for such robots to assemble one product can be represented by variables D, S, I, and the drilling task takes n_D time units and the screwing task takes n_S time units. The following constraints can be straightforwardly obtained:

$$D < S < I$$
$$S = D + n_D$$
$$I = S + n_S$$

Constraints can be further defined by their *extension* in terms of logic operation formulas.

Example: In earlier example about the starting time of Monday classes, each class has a duration of 75 minutes, except the class starting at 17 has 150 minutes. Professor p_1 has classes starting at 8 and 15:30; professor p_2 has classes starting at 11 and 15:30; professor p_3 has classes starting at 14; professor p_4 has classes starting at 8 and 17; professor p_5 has classes starting at 9:30 and 11:30. The department chair wants to organize a meeting for 2 hours and 15 minutes, with possible time slots 8:30–10:45, 10–12:15, 12–14:15, and 14:30–16:45. How can we find the best possible meeting time slot(s) of most number of professors, and more than half professors, to attend? Of course, the method should be appropriate for computation.

Solution: We can first define variable A as the class starting at 8, and subsequently for $B, \cdots, G$. We further define $\psi_i, i = 1, \cdots, 5$ as the time duration that is not available for professor p_i. Then,

$$\psi_1 = A \wedge F$$
$$\psi_2 = C \wedge F$$
$$\psi_3 = E$$

$$\psi_4 = A \wedge G$$
$$\psi_5 = B \wedge D$$

For the time slot 8:30–10:45 (i.e. T_1 that includes class duration $A \vee B$, we can set the indicator functions can be defined as $\mathbb{I}_i = \neg [\psi_i \wedge (A \vee B)]$, $i = 1, \cdots, 5$ to indicate whether p_1 is available in this time slot. Then, the number of professors who can attend the meeting at time slot T_1 is

$$N_{T_1} = \sum_{i=1}^{5} \mathbb{I}_i \tag{1.5}$$

Similarly, we can get the results for all time slots, and only T_2 10–12:15 is a good choice since only $N_{T_2} \geq 3$. This constraint problem can be thus solved by the logic operations, which serves the simplest form of AI problem. ■

Above example is very common in our daily life but actually more complicated than it appearing if exploring computational complexity. A popular way to resolve above example is to set up a questionnaire for every potential participant, like what doodle.com is doing. The host (i.e. department chair in the example) sets up possible time slots and each participant to reply his/her availability, then the computer just finds the statistics by counting to reach conclusion. Please note that above solution precisely follows this approach, except replying availability by each potential participant is usually done by human intelligence while above solution executes in terms of logic operations.

1.3.1 Constraint Satisfaction Problems

The scheduling example is actually viewed as a *constraint satisfaction problem* (CSP) consisting of

- a set of variables
- a domain for each variable
- a set of constraints

We can use the following example to comprehend CSP.

Example (Delivery Robot): A delivery robot delivers items a, b, c, d, e and each delivery activity may take place at one of the time instants t_1, t_2, t_3, t_4. Denote A to be the variable representing the time instant that delivery activity for a occurs, and so on. The domains of variables A, B, C, D, E are

$$dom(A) = \{t_1, t_2, t_3, t_4\}$$

$$dom(B) = \{t_1, t_2, t_3, t_4\}$$
$$dom(C) = \{t_1, t_2, t_3, t_4\}$$
$$dom(D) = \{t_1, t_2, t_3, t_4\}$$
$$dom(E) = \{t_1, t_2, t_3, t_4\}$$

Suppose we have the following constraints to be satisfied:

$$\{(B \neq 1), (C \neq 4), (A = D), (E < B), (E < D), (C < A),$$
$$(B \neq C), (A \neq E)\}$$

▶ **Exercise:** Given above set up, please answer the following questions:

(a) Is there a model?
(b) If so, find a model.
(c) How many models can we have?
(d) Please find all the models.
(e) What is the best model if appropriate performance measure is supplied? One possible performance measure is the minimal number of required time instances.

The *assignment space*, $\mathfrak{D}$, denotes the set of all assignments. In the example of delivery robot, the assignment space is

$$\mathfrak{D} = \{(A = 1, B = 1, C = 1, D = 1), (A = 1, B = 1, C = 1, D = 2), \cdots ,$$
$$\cdots , (A = 4, B = 4, C = 4, D = 4)\}$$

A finite CSP can be intuitively solved by the exhaustive *generate-and-test algorithms*. For the example of delivery robot, there are $|\mathfrak{D}| = 4^5 = 1,024$ different assignments to be tested, which exponentially grows with the number of delivery items and is clearly not scalable in computations. If each of the m variable domains has the size d, $\mathfrak{D}$ has d^m elements to test. If there are c constraints, the total number of exhaustive tests is $O(cd^m)$, which can easily turn to computationally impossible. An intelligent method is therefore required.

Typically, the most straightforward approach is to conduct an efficient search to solve the problem, while *search* is generally considered as a core technology in artificial intelligence. In particular, CSP consisting of linear constraint functions form a sort of *linear programming* (LP) problems.

Linear programming was first systematically studied by the Russian mathematician Leonid Kantorovich in 1939. For a LP problem, the constraints must be linear inequalities forming a convex set and the objective function be linear too. The time complexity of linear programming is polynomial in the number of variables. Linear programming is probably the most widely studied and broadly useful class of optimization problems. It is a special case of the more general problem of *convex optimization*, in which the constraint region is any convex region and the objective function is also convex within the constraint region. Under certain conditions, convex optimization problems are also polynomially solvable and may be feasible in practice even with thousands of variables.

Example (Coloring Map): There are 48 states in the mainland United States. We want to paint a color for each state and the neighboring states must be painted in different color. This is again a CSP.

There is a deeper version of the coloring map problem, that is, what are the minimum number of colors to paint any map, which is know as *four-color problem* in *graph theory*. Graphs provide a useful means to comprehend a lot of AI problems related to robotics.

1.3.2 Solving CSP by Search

Back to efficiently solve CSP, we may first note:

(a) As each constrain only involves a subset of the variables, some constraints can be tested before all the variables have been assigned values.

(b) As long as a partial assignment is inconsistent with a single constraint, any total assignment that extends this partial assignment is also inconsistent.

In example of delivery robot, the assignment $(A = 1, B = 1, C = 1, D = 1)$ is inconsistent with the constraint $B \neq C$. Such an inconsistency can be discovered prior to any value assignment and thus save the amount of computation.

Similar to the concept of state transition diagram introduced earlier, we can leverage a mathematical tool, *graph theory*, to more systematically examine the general class of such problems.

Box (Basics in Graphs)

The relationship among variables can be mathematically described by a *graph*. A graph that is a collection of *vertices*, $\mathcal{V}$, joined by a collection of *edges*, $\mathcal{E}$. Mathematically, we can write $\mathcal{G} = (\mathcal{V}, \mathcal{E})$. In a graph $\mathcal{G}$, the number of vertices is called the *order* of $\mathcal{G}$ and the number of edges is known as its *size*. The number of edges of a node is known as the *degree* of this node.

Theorem (Handshaking Lemma): Let $\mathcal{G}$ be a graph of order n and size m with vertices $v_1, \cdots, v_n$. Then

$$deg(v_1) + \cdots + deg(v_n) = 2m \qquad (1.6)$$

A particular graph is known as *tree*, which is very useful to develop efficient algorithms.

Definition: A connected graph that contains no cycle is called a *tree*. A vertex of degree 1 in a tree is usually called as a *leaf*.

Remark: A leaf can be considered as a terminal node. The root node can be considered as the original or starting point/node.

Theorem: A graph G is a tree if and only if every two vertices of G are connected by only one path.

Theorem: Every tree of at least two vertices has at least two leaves.

Theorem: Every tree with n vertices has $n - 1$ edges.

Corollary: If T is a tree of order n and size m whose vertices are $v_1, v_2, \cdots, v_n$, then

$$deg(v_1) + deg(v_2) + \cdots + deg(v_n) = 2m = 2(n - 1) \qquad (1.7)$$

A more computationally effective method is to construct a search space and then to employ an appropriate search algorithm. A typical way to construct the search space is to establish a graph, which is useful in the delivery robot example. The graph search is defined as follows:

- The start node is the empty assignment (i.e. no assigned value to any variable).
- The nodes represent assignments of values to some subset of the variables.

- A goal node is a node of an assigned value to every variable, and such assignment is consistent with all constraints.
- The neighbors of a node n are obtained by selecting a variable Y that is not assigned in node n and each assignment of a value to Y does not violate any constraint. In other words, suppose node n is the assignment $\{V_1 = v_1, \cdots, V_k = v_k\}$. In order to find the neighbors of node n, select $U \notin \{V_1, \cdots, V_k\}$. $\forall y \in dom(Y)$, $\{V_1 = v_1, \cdots, V_k = v_k, Y = y\}$ does not violate (i.e. is consistent) with all the constraints, and then the node $\{V_1 = v_1, \cdots, V_k = v_k, Y = y\}$ is a neighbor of node n.

Box (Search in A Directed Graph)

We are also interested in the *directed graphs*. A directed graph consists of a set of nodes $\mathcal{N}$ and a set of arcs $\mathcal{A}$, where an *arc* is an ordered pair of nodes. The arc $\langle n_1, n_2 \rangle$ is an *outgoing arc* from n_1 and an *incoming arc* to n_2. The arc $\langle n_1, n_2 \rangle$ also indicates n_2 is a neighbor of n_1, but the reverse is not necessarily true.

A *path* from node s to node t is a sequence of nodes $\langle n_0, n_1, \cdots, n_k \rangle$ such that $s = n_0$ and $t = n_k$, or a sequence of arcs, $\langle n_0, n_1 \rangle$, $\langle n_1, n_2 \rangle, \cdots, \langle n_{k-1}, n_k \rangle$. A *goal* is a Boolean function on nodes. If $goal(n)$ is true, then the node n satisfies the goal and n is the goal node.

Figure 1.11 illustrates a general case of graph search. Efficient search over a graph plays a central problem for AI. More fundamental search algorithms will be introduced in next chapter.

In many cases, we are interested in the goal node(s) rather than the paths from the start node in search of solution(s). We can use constraints to remove possibilities of nodes or paths such that solutions can be obtained.

Example (Delivery Robot Continued): Following the example of Delivery Robot, we construct a sub-tree for $A = 2$ by considering the constraints denoted by the cross signs in Figure 1.12.

- $B \neq 1$: dark red cross
- $C \neq 4$: red cross
- $C < A$: green cross
- $A = D$: orange cross
- $E < D$: yellow cross

We therefore search according to the depth, in the order of $A \rightarrow B \rightarrow C \rightarrow D \rightarrow E$, which is an example of *depth-first search*. We will look into further in later Chapter.

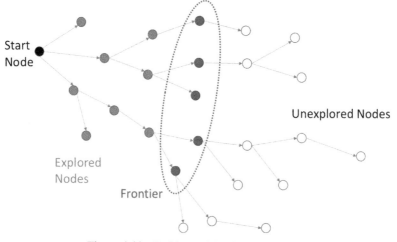

Figure 1.11 Problem solving by graph search.

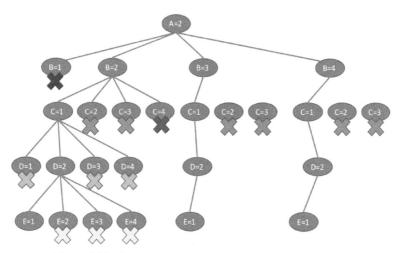

Figure 1.12 Partial illustration of search over the tree for CSP.

▶ **Exercise:** In the example of Delivery Robot,

(a) How many logic operations of searching the entire tree for solutions (i.e. generate-and-test)?

(b) As the illustration of Figure 1.12, for $B \neq 1$, we can prune the entire sub-tree. By such pruning techniques, how many logic operations of tree searching are really required?

Search the tree with depth-first search is typically known as *backtracking*, which is surely more efficient than generate-and-test. We will discuss further in Chapter 2.

The consistency algorithms can be efficiently operating over the network of constraints for by the CSP by the following principles, and such a network is called a *constraint network*.

- Each variable is represented as a node, say drawn in a circle.
- Each constraint is represented as a node, say drawn in a rectangular.
- A set D_X of possible values is associated with each variable X.
- For every constraint c, and for every variable X in the scope of c, there is an edge $\langle X, c \rangle$.

Box (Order of Complexity)

Let $T(n)$ be a function, say the worst running time of a certain algorithm on an input variable of size n. Given another function $f(n)$, $T(n)$ has the order of $f(n)$, denoted as $O(f(n))$, if the function $T(n)$ is bounded above by a constant multiple of $f(n)$ for sufficiently large n. That is, $T(n)$ is $O(f(n))$ if there exists constants $c > 0, n_0 \geq 0$, such that $\forall n \geq n_0$, we have $T(n) \leq c \cdot f(n)$, or equivalently say T is a asymptotically upper-bounded by f.

Similarly, we say $T(n)$ is $\Omega(f(n))$ if there exist constants $\epsilon > 0, n_0 \geq 0$, such that $\forall n \geq n_0$, we have $T(n) \geq \epsilon \cdot f(n)$, or equivalently say T is asymptotically lower-bounded by f.

If a function $T(n)$ is both $O(f(n))$ and $\Omega(f(n))$, then $T(n)$ is $\Theta(f(n))$, which means that $f(n)$ is an asymptotically tight bound for $T(n)$.

Example (delivery robot continued): The constraint network of robot delivery problem can be drawn as Figure 1.13.

In the simple case, when a constraint has just one variable in its scope, the arc (or edge) is *domain consistent* if every value of the variable satisfies the constraint. Suppose constraint c has the scope $\{X, Y_1, \cdots, Y_k\}$. $\langle X, c \rangle$ is *arc consistent* if $\forall x \in D_x$, there are values $y_1, \cdots, y_k$ and $y_i \in D_y$, such that $c(X = x, Y_1 = y_1, \cdots, Y_k = y_k)$ is satisfied. If all arcs in a network are arc consistent, the network is arc consistent. Home Project 1 will give one problem to apply above knowledge.

▶ **Exercise (Coloring a Map):** Please color the 16 states in the south of the US, while neighboring states must be painted in different colors. The colors are used in the following order (i) green (ii) yellow (iii) blue (iv) red (v) black.

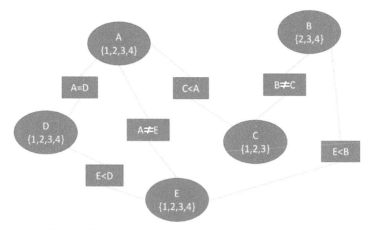

Figure 1.13 Constraint network for robot delivery problem.

Figure 1.14 16 States in the south of US (excluding washington DC).

(a) Please develop the model of the problem and then an algorithm to color the south states in the US. What is the minimum number of colors?
(b) Please calculate the number of search step in your algorithm? Please find the required memory space in your algorithm?
(c) Please develop the algorithm and program to paint 48 states in 4 colors.

Further Reading: [2] provides detailed and holistic introduction of artificial intelligence, while the first two chapters are excellent further reading at this point. [3] supplies comprehensive knowledge in algorithms and introduction on graphs and graphical algorithms.

References

[1] E. Ko, K.-C. Chen, "Wireless Communications Meets Artificial Intelligence: An Illustration by Autonomous Vehicles on Manhattan Streets", *IEEE Globecom*, 2018.

[2] Stuart Russell, Peter Norvig, *Artificial Intelligence: A Modern Approach*, 3rd edition, Prentice-Hall, 2010.

[3] J. Kleinberg, E. Tardos, *Algorithm Design*, 2nd. Edition, Pearson Education, 2011.

2

Basic Search Algorithms

From Chapter 1, we may note that a core issue to design an AI agent is the capability to search the solution(s) achieving its goal, where such agents are known as *goal-based agents*. Most robots have their goals or missions. Agents of (artificial) intelligence are supposed to maximize their performance measure, which can be simplified as an agent adopting a goal and aiming to satisfy this goal. The simplest task environment whose solution is therefore a sequence of actions. However, in general cases, the agent's future actions may depend on future percepts. The process that is looking for a sequence of actions to reach the goal is called *search*. In this chapter, some basic search algorithms will be introduced.

2.1 Problem-Solving Agents

A problem solving agent is a kind of goal-based agents, and uses *atomic* representations where the states of the world are considered as wholes, with no internal structure visible to the desired problem-solving algorithms. Goal-based agents can use more sophisticated structure of representation to be named as *planning agents*, while planning will be discussed in later chapter.

▶**Example (Touring):** Jeremy's family living in Boston plans a winter vacation trip to visit a few cities in Florida, including Tallahassee, Jacksonville, Orlando, Tampa, and Miami. They will fly to one of these cities, rent a car driving to these cities via inter-state highways (i.e. I-95, I-75, I-10, and I-4), and then fly out from the flying-in city. Please determine (1) which city to fly in (2) what is the best route to drive, while the optimal (i.e. best) route is measured by the distance.

Solution: The most common way to solve this problem is to construct a graph corresponding to the problem, as shown in Figure 2.1.

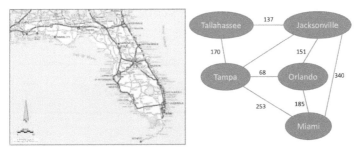

Figure 2.1   Touring in Florida.

A *problem* can be formally defined by five components:

- The *initial state* in which the agent starts from.
- A description of the possible *actions* available to the agent. Given a particular state s, ACTIONS(s) returns the set of actions that can be executed in s, which means that each of these actions is applicable in s.
- A description regarding what each action does, which is known as the *transition model*, specified by a function RESULT(s, a) that returns the state that results from doing action a in state s. The term *successor* is referred to any state reachable from a given state by a single action. For any specific problem, the initial state, actions, and transition model implicitly define the *state space*, which is the set of all states reachable from the initial state by any sequence of actions. The state space forms a *directed network* or *graph* in which the nodes are states and the links (or edges) between nodes are actions. A *path* in the state space corresponds to a sequence of states connected by a sequence of actions.
- The *goal test*, which determines whether a given state is a *goal state*. Please note that there can be multiple goal states.
- A *path cost* function that maps a numeric cost to each path. The problem-solving agent chooses a cost function that reflects its own performance measure.

The preceding elements define a *problem* and can be gathered into a single data structure as the input to a problem-solving algorithm. A *solution* to a problem is a sequence of actions that lead from the initial state to a goal state. The quality of solution is measured by the path cost function, and an *optimal solution* has the lowest path cost among all solutions.

In order to deal with real-world problems, *abstraction* that removes some details from representation, which is useful if carrying out each of the actions in the solution is easier than the original real-world problem. Consequently,

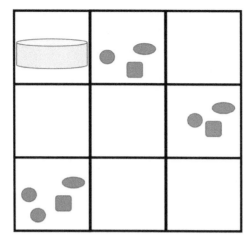

Figure 2.2 The silver cylinder represents a cleaning robot to sense first and then decide to clean for 8 tiles.

the abstraction is valid if we can expand the corresponding abstract solution into a solution in the more detailed world. The subsequent abstract solution corresponds to a large number of more detailed paths.

Example: Let us look into a em toy problem as Figure 2.2. Standing on one square tile, a cleaning robot examines all 9 square tiles, one-by-one, by first sensing the existence of dirts and then taking an action to clean, or moving to next tile.

In this toy example, a cleaning robot is the agent of our interest, and we can formulate a problem as follows.

States: The states are defined by the agent location and the locations of dirt. There are n possible locations for the agent, while 9 locations in this example. Each location may or may not contain dirt. Therefore, there are totally $n \cdot 2^n$ states.

Initial State: Left-upper tile represents the initial state.

Actions: The agent in each state has possible actions, cleaning, and moving to next location with four directions (up, down, right, left).

Transition model: The actions result in expected efforts. For example, cleaning in a square tile has no effort, except prohibited moving (e.g. moving left in the leftmost square tiles.

Goal test: It checks whether all square tiles are clean.

Path cost: Since each movement and cleaning action consumes energy, the cost is the number of steps and cleaning actions.

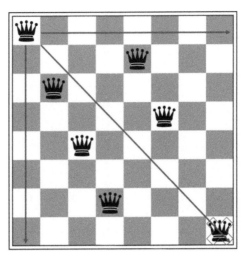

Figure 2.3   An illustration to search a solution of eight-queen problem, where the red cross indicates the violation with the leftmost-uppermost queen.

Example (Eight-Queen Problem): The well known eight-queen problem is to place 8 queens on the chess board without violation as each queen can move horizontally, vertically, and diagonally. This problem can be formulated as a typical search problem.

The 8-queen problem can be generally formulated in two ways: incremental formulation and complete-state formulation. An incremental formulation involves operators that augment the state description, starting from an empty state. For the 8-queens problem, each action adds a queen to the state until reach the goal. A complete-state formulation starts with all 8 queens on the board and moves them around until obtaining the solution. An immediate incremental formulation is as follows:

States: Any arrangement from 0 to 8 queens on the board.

Initial state: 0 queen on the board.

Actions: Add a queen to an empty square (i.e. no violations with previously placed queens).

Transition model: Return the board with a queen added to the specified square.

Goal test: 8 queens on the board without any violation.

The complexity of this immediate formulation suggests $64 \cdot 63 \cdots 57 \approx 1.8 \times 10^{14}$ possible sequences of actions to search, which is obviously inefficient. A straightforward improvement can greatly reduce the number of state spaces to 2,057 by the following modified formulation:

States: Possible placements of $n, 0 \leq n \leq 8$ queens are arranged by one per column in the leftmost columns.

Actions: Add a queen to any square in the leftmost empty column such that no violation occurs.

▶ **Exercise (Eight-Queen Problem):** For the 8-queen problem,

(a) Please develop an algorithm to find out solutions, and calculate the number of states in your search algorithm.

(b) Considering $90°$ rotationally invariant, how many distinct solutions can you find?

2.2 Searching for Solutions

As indicated in the Chapter 1, an AI problem can be solved by constructing a graph of tree and searching over the tree for possible solution(s). More precisely, since a solution is a sequence of actions, search algorithms work by considering various possible action sequences. The possible action sequences starting at the initial state form a *search tree* with the initial state at the *root*, and the *branches* represent actions and the *nodes* correspond to states in the state space of the problem.

Example (Touring, Continued): We can use tree generation to find the solutions of Touring in Florida problem.

Solution: The first step is to select a city to fly in and then a search tree can be constructed. Such a search tree can be viewed as a sub-tree of the entire search tree. in other words, we expand the current state by applying each legitimate action to the current state, thereby generating a new set of states. Then, we add tree branches from the *parent node*, which lead to new *child nodes*.

The Figure 2.4 illustrates a portion of generate sub-tree starting from Orlando (i.e. flying in Orlando). The green color node indicates returning to the root of sub-tree and satisfy the goal test, which suggests such a path is a solution but might not be an optimal solution. In summary for this solution, fly in Orlando, then drive to Tampa, to Tallahassee, to Jacksonville, to Miami, and back to Orlando.

The same procedure can be used to generate other sub-trees by designating a city to fly in. Then, we obtain the entire search tree for this touring problem.

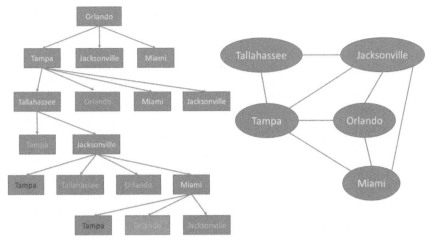

Figure 2.4   A portion of sub-tree starting from Tampa in the Florida touring problem, while the graphical representation of inter-state highway paths at the right.

In Figure 2.4, the nodes in orange color indicate returning to a node that departs earlier, a *repeated state* that might indicate inefficiency. The red color wording nodes indicate a *loop path*, which implies potentially disfavored situation forming an infinite loop.

Loopy paths are a special case of the more general concept of redundant paths, which exist whenever there is more than one way to get from one state to another. To avoid exploring the redundant paths, remembering the history of exploration appears a good way by amending a data structure called *explored set*.

▶ **Exercise (Touring):** Please find all possible solutions by complete the tree generation in Figure 2.4. How many solutions do you get?

Remark: The famous Traveling Salesman problem (TSP) is very similar to the Touring problem. TSP is a touring problem in which each city must be visited exactly once. The aim is to find the shortest tour. The problem is known to be *NP-hard*, but an enormous amount of effort has been expended to improve the capabilities of TSP algorithms due to its wide applications to networking routing, parallel processing, finance, and various management topics.

Figure 2.5 summarizes the algorithms for tree search and graph search, where the bold Italic lines are designed to handle repeated states.

Algorithm: Tree-Search

Input: Root of the tree r
Output: goal, return *null* if goal is not found
1 frontier ← a set containing r
2 goal ← *null*
3 **while** frontier *is not empty* **do**
4   node v ← frontier.*remove*
5   **if** v = goal **then**
6   goal ← v
7   **break**
8   **else**
9   frontier ← v.*children*
10 **return** goal

Algorithm: Graph Search

Input: Starting node u
Output: goal, return *null* if goal is not found
1 frontier ← a set containing u
2 explored ← an empty set
3 goal ← *null*
4 **while** frontier *is not empty* **do**
5   node v ← frontier.*remove*
6   **if** v = goal **then**
7   goal ← v
8   **break**
9   **else**
10   explored ← v
11   **for** *each node w in v.neighbors* **do**
12   **if** w *not in* explored **then**
13   explored ← w
14 **return** goal

Figure 2.5 Tree search and graph search algorithms, from reference [3].

Each search algorithm requires a data structure to keep track of the search tree being constructed. For each node n of the tree, it is associated with a structure of the following four components:

- n.STATE: the corresponding state for node n in the state space
- n.PARENT: the node that generates node n in the search tree
- n.ACTION: the action applied to the parent node to generate node n
- n.PATH-COST: the cost, traditionally denoted by $g(n)$, associated with the path from the initial state to node n, which can be indicated by the parent pointer(s) in programming.

Given the components of a parent node, it is straightforward to compute the components for any child node. Consequently, we can establish a function

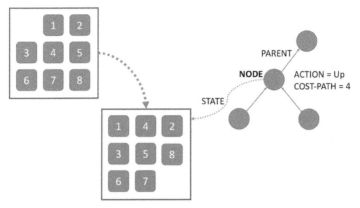

Figure 2.6 Data structure of shuffle game.

CHILD-NODE, which takes a parent node and an action to return the resulting child node.

Remark: A node is a bookkeeping data structure used to represent the search tree. A state corresponds to a configuration of the world.

Example: Each block in the shuffle game can move Up, Down, Left, Right, as long as there exists a space. The typical game is from a state of unknown moves back to the original state. The magic cube (or known as Rubik's cube) can be viewed as an example extended from this game. Figure 2.6 indicates the situation of 4 moves (cost-path=4), and the 4th move is Up. In this figure, nodes are associated with the data structures from the search tree. Each node has a parent, a state, and bookkeeping field(s). An arrow points from a child to the parent.

In a tree search, the frontier shall be stored such that the search algorithm can easily select the next node to expand accordingly. The appropriate data structure to this scenario is a *queue*, which supports operations of Empty, Pop, and Insert.

Box (Queue):

A *queue* is a kind of data structure that is characterized by the order in which they store the inserted data elements (say, nodes for a graph).

Three common variants are the *first-in-first-out* (FIFO) queue, which pops the oldest element of the queue; the *last-in-first-out* (LIFO) queue that is also known as a *stack*, which pops the newest element of the queue; and the *priority queue*, which pops the element of the queue with the highest priority according to some ordering function.

► **Exercise:** Please establish the search tree for the scenario of Figure 2.6, write down algorithm and program to execute, so as to reach back to original state.

Another critical issue is to select appropriate search algorithm. We usually evaluate the performance of an algorithm by considering:

- *Completeness* to guarantee the existence of a solution
- *Optimality* to find an optimal solution
- *Time complexity*, the computing time/complexity
- *Space complexity*, required memory to execute.

► **Exercise:** (Graphical Algorithm) Serena wants to host a party and to determine whom to be invited. She has n persons to choose from, and she has made up a list for the pairs of these n persons knowing each other (i.e. not necessarily all know each other). She wants to invite as many as possible persons, subject to the constraints: for each person attending the party (i) at least $k \ll n$ other persons who know (ii) at least k persons in the party who do not know.

(a) Please develop the mathematical model to solve the problem.
(b) Please develop the algorithm which efficiently solve the problem.

2.3 Uniform Search

The search strategies have no further information beyond what have been provided in the problem are categorized as *uniform search* or *blind search*. Uniform search just generates successors and distinguishes between the goal state and non-goal states. All uniform search strategies are distinguished by the order in which nodes are expanded. For search strategies knowing whether one no-goal state is more promising than another, they are called *informed search* or *heuristic search*.

2.3.1 Breadth-First Search

Breadth-first search is a simple and intuitive strategy, in which the root node is expanded first, then all the successors of the root node are expanded next, then their successors, and repeating the procedure. In principle, all the nodes are expanded at a given depth in the search tree prior to expanding to next level. Breadth-first search can be easily implemented by a FIFO queue for the frontier. As a consequence, new nodes (which are always deeper than their parents) go to the back of the queue, while old nodes (which are shallower than the new nodes) get expanded first. Slightly different from the general

---

Algorithm: Breadth-First Search

Input: Starting node u
Output: goal, return *null* if goal is not found
1 frontier ← a FIFO queue containing u
2 explored ← an empty set
3 goal ← *null*
4 **while** frontier *is not empty* **do**
5 │ node v ← frontier.*dequeue*
6 │ if v = goal **then**
7 │ │ goal ← v
8 │ │ **break**
9 │ **else**
10 │ │ explored ← v
11 │ │ **for** *each node w in v.neighbors* **do**
12 │ │ │ if w *not in* explored **then**
13 │ │ │ │ explored ← w

14 **return** goal

Figure 2.7   Breadth-first search on a tree from [3].

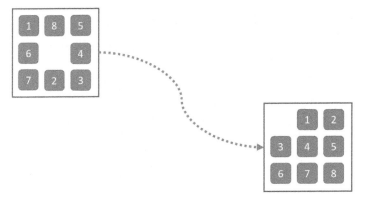

Figure 2.8   Recovery of shuffled blocks.

graph-search algorithms, breadth-first search executes goal test when a node is generated, rather than being selected for expansion.

▶ **Exercise:** Figure 2.8 shows the blocks have been shuffled for a number of times. Please develop an algorithm to recover the original version. Please evaluate the (time) complexity of your algorithm.

Please recall the four criterion to evaluate an algorithm. How about the breadth-first search?

- It is obviously complete, since the goal node will be eventually found as long as the tree is finite.

- However, the shallowest goal node is not necessarily optimal. As a matter of fact, breadth-first search is optimal if the path cost of a node is a nondecreasing function of the depth of this node. The most common such scenario, for example, is that all actions have the same cost.
- The (time) complexity is unfortunately not good. For the simple case of searching a uniform tree that each node has b branching successors (i.e. child nodes), at the depth of d, the worst case scenario searches the total number of nodes $b + b^2 + \cdots + b^d$, and results in time complexity $O(b^{d+1})$.
- Regarding space complexity, for any kind of graph search, it shall store every expanded node in the explored set. For breadth-first graph search, in particular, every node generated in the frontier remains in memory. The consequent space complexity of breadth-first search is $O(b^d)$.

Remark: The challenge of breadth-first search lies in memory complexity, in addition to time complexity. Actually, any search problem of exponential complexity cannot be solved by uninformed methods, except for small dimensions.

When the costs of all steps are equal, breadth-first search is effective because it always expands the shallowest unexpanded node. By a simple extension, we can develop an optimal algorithm with any step-cost function. Instead of expanding the shallowest node, *uniform-cost search* expands the node n with the lowest path cost $g(n)$, which can be implemented by storing the frontier as a priority queue ordered by g. In addition to the ordering of the queue by path cost, there are two significant aspects different from the breadth-first search. The first difference lies in the goal test being executed at a node that is selected for expansion rather than when it is first generated. The reason is that the first goal node that is generated may be on a suboptimal path. The second different aspect is an amended test when a better path is found to a node currently on the frontier.

In other words, the uniform-cost search algorithm is identical to the general graph search algorithm as Figure 2.9, except the use of a priority queue and the addition of an extra check when a shorter path to a frontier state is discovered. The data structure for the frontier needs to support efficient membership testing, so it should combine the capabilities of a priority queue and a table in implementation.

▶ **Exercise:** Please use breadth-first search and uniform-cost search to solve the Florida touring problem as Figure 2.1, and then compare the complexity.

Algorithm: Uniform-Cost Search

Input: Starting node u
Output: goal, return *null* if goal is not found
1 frontier $\leftarrow$ a priority queue ordered by Path-Cost, containing u
2 explored $\leftarrow$ an empty set
3 goal $\leftarrow$ *null*
4 **while** frontier *is not empty* **do**
5   node $v \leftarrow$ frontier.*dequeue*
6   **if** v = goal **then**
7   goal $\leftarrow v$
8   **break**
9   **else**
10   explored $\leftarrow v$
11   **for** *each node w in v.neighbors* **do**
12   **if** w *not in* explored **then**
13   explored $\leftarrow w$

14 **return** goal

Figure 2.9 Uniform cost search from [3].

2.3.2 Dynamic Programming

The method of *dynamic programming*, originally coined by Bellman in 1957 for the purpose of control but naming in computing terminology, systematically develops solutions for all subproblems of increasing costs or metrics (such as lengths), can be seen as a special format of breadth-first search on graphs.

The mathematical formulation of dynamic programming can be understood as the decision-control mechanism of a dynamic system. The outcome of each action (i.e. decision) is not fully predictable using *a priori* probability, but can be observed prior to when the next action will be made, with the objective to minimize the cost or any metric (e.g. length).

The basic problem is formulated by two features:

(a) an underlying discrete-time dynamic system
(b) a metric/cost functional that is additive over the time.

The dynamic system is represented by

$$x_{k+1} = f_k(x_k, a_k, w_k), \quad k = 0, 1, \cdots, K \tag{2.1}$$

where

    k: the discrete time index
    x_k: the state of the dynamic system

a_k: action (or decision) variable to be taken at the time k with the
   knowledge of state x_k, as a means of control
w_k: independent random variables from observation, say noise
K: the horizon of action
The cost functional is additive in the sense that

$$c_k(x_k, a_k, w_k) = c_K(x_K) + \sum_{k=1}^{K-1} c_k(x_k, a_k, w_k) \qquad (2.2)$$

where $c_K(x_K)$ is the terminal cost incurred at the end of decision process.
We intend to formulate such a problem of selecting the sequence of actions
$a_0, a_1, \cdots, a_{K-1}$ such that the following expected cost is maximized

$$\mathbb{E}\left[c_K(x_K) + \sum_{k=1}^{K-1} c_k(x_k, a_k, w_k)\right] \qquad (2.3)$$

Example (Inventory Management): We can formulate the inventory
management as a dynamic programming problem. Let x_k represent the stock
in (time) episode k, a_k be the stock ordered/delivered in episode k, and w_k
stand for the demand in episode k given the probabilistic or empirical distribution. Assuming $H(\cdot)$ as the holding cost of state and c as the unit cost of
delivered stock, this inventory system can be described as Figure 2.10, where

$$x_{k+1} = x_k + a_k - w_k$$
$$c_k = c \cdot a_k + H(x_{k+1})$$

Considering a dynamic system of a finite state space $\mathcal{S}_k$, at any state x_k, an
action/decision a_k (serving the purpose of control) is associated with a state
transition from x_k to $f_k(x_k, a_k)$. Such a dynamic system can be therefore

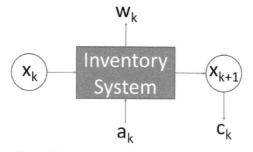

Figure 2.10   Modeling of an inventory system.

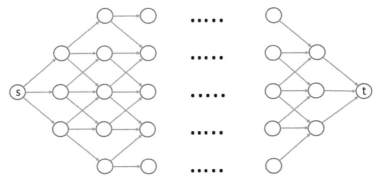

Stage 0   Stage 1   Stage 2   Stage K-1   Stage K

Figure 2.11   State transition diagram, with s as the initial state (starting node) and t as the artificial terminal node.

represented as a graph illustrated in Figure 2.11, where each (directed) edge of a graph is associated with a cost. We amend an artificial terminal node t. Each edge connecting a state x_K at stage K to the terminal mode has the associated cost $c_K(x_K)$. The decisions corresponding to paths originating from the initial state (i.e. node s) and terminating at a node corresponding to the final stage k. Given the cost associated with each edge, the policy of decisions/actions is equivalent to finding the shortest path from node s to node t.

Proposition (Dynamic Programming):

By defining

c_{ij}^k: cost of transition from state $i \in \mathcal{S}_k$ to state $j \in \mathcal{S}_{k+1}$, $k = 0, 1, \cdots, K - 1$

c_{it}^K: terminal cost of state $i \in \mathcal{S}_K$

dynamic programming is formed as

$$J_K(i) = c_{it}^K, i \in \mathcal{S}_K \tag{2.4}$$

$$J_k(i) = \min_{j \in \mathcal{S}_{k+1}} \left[c_{ij}^k + J_{k+1}(j) \right] \tag{2.5}$$

Recall that the sequence of actions (or decisions) is known as a *policy*. The optimal cost of the policy is $J_0(s)$, equal to the weighted length of the shortest path from s to t. Please note that above version of dynamic programming proceeds backward in time, while an equivalent forward alternative is possible.

We may apply dynamic programming to search the policies. Therefore, the 8-queen problem and the touring problem can be solved via dynamic programming by identifying the optimal policy.

▶ **Exercise:** Suppose we have a machine that is either functioning or broken down. If it functions for an entire day, a gross profit $200 will be generated. If the machine is broken, it generates zero profit. The probability for this machine goes broken in a day is 0.3 if high-quality material has been used. The probability for this machine goes broken in a day is 0.7 if ordinary material has been used. The extra cost for high-quality material is $20 per day. When this machine breaks down at the beginning of a day, (a) it can be repaired at the cost of $50, with probability of failure 0.2 on the same day (b) it can be replaced with the cost $150 to guarantee running for the day. Assuming a new machine at the starting of the first day, please find the optimal policy of repair, replacement, and (high-quality) material, to maximize total profit over 7 days.

Dynamic programming is useful to a general class of problems, *shortest path* problems. A major application scenario of the shortest path algorithm is to find the route from the source node to destination node in any connected graph or network, as long as the distance associated with each link/edge in the network/graph is well defined. Usually, each link is assigned a positive number that can be viewed as the length or the cost for execution over this link. Ideally, such a measure (i.e. length/distance or cost) is a metric, but not always due to possible asymmetric distance measure, that is, a link may have different lengths or costs in two possible directions. Each path between these two nodes has a total length equal to the sum of the lengths (i.e. cost) of links going through. A shorted path routing algorithm identifies the path(s) of minimal end-to-end length between source and destination.

An important distributed algorithm for calculating shortest paths to a given destination, known as the *Bellman-Ford algorithm* has the form

$$D_i = \min_j(d_{ij} + D_j) \tag{2.6}$$

where D_i is the estimated shortest distance from node i to the destination and d_{ij} is the length/cost of link/arc $\langle i, j \rangle$.

Practical implementation of *Bellman-Ford algorithm* supposes the destination node as node 1 and considers the problem of finding the shortest path from each node to node 1. We assume that there exists at least one path from every node to the destination. Denote $d_{ij} = \infty$ if $\langle i, j \rangle$ is not an arc of the graph. A shortest walk from mode i to node 1, subject to the the constraint that the walk contains at most h arcs and goes through node 1 only once, is referred to as the shortest walk and its length is denoted by D_i^h. D_i^h can be generated by the iteration:

$$D_i^{h+1} = \min_j(d_{ij} + D_j^h), \forall i \neq 1 \tag{2.7}$$

with initial conditions $D_i^0 = \infty, \forall i \neq 1$ and terminating condition after h iterations $D_i^h = D_i^{h+1}, \forall i$.

Proposition (Bellman-Ford Algorithm):

Consider the Bellman-Ford algorithm with the initial conditions $D_i^0 = \infty$, $\forall i \neq 1$. Then,

(a) D_i^h generated by the algorithm using Equation (2.7) represent the shortest lengths of walk from node i to node 1 in no more than h arcs.

(b) The algorithm terminates after a finite number of iterations if and only if all cycles not containing node 1 have nonnegative length. Furthermore, if the algorithm terminates, it does so after at most $h \leq H$ iterations, and at termination, D_i^h is the shortest path length from node i to node 1.

Intuitively, the Bellman-Ford algorithm first finds the one-arc shortest walk lengths, then two-arc ones, and so forth. We can also argue that the shortest walk lengths are equal to the shortest path lengths, under the additional assumption that all cycles not containing node 1 have nonnegative length. Another shortest path algorithm requiring non-negative cost/length of any arc is *Dijkstra algorithm*, whose worst-case computational requirements are considerably less than those of Bellman-Ford algorithm. The general idea is to find the shortest paths in order of increasing path length. The shortest of the shortest paths to node 1 must be the single-arc path from the closest neighbor of node 1, since any multiple-arc path cannot be shorter than the first arc length because of the nonnegative-length assumption. The next shortest of the shortest paths must either be the single-arc path from the next closest neighbor of 1 or the shortest two-arc path through the previously chosen node, and so on.

We can view each node i as being labeled with an estimate D_i of the shortest path length to node 1. When the estimate becomes certain, we regard the node as being permanently labeled and keep track of this with a set of permanently labeled nodes, $\mathcal{P}$. The node added to $\mathcal{P}$ at each step will be the closest to node 1 out of those that are not yet in $\mathcal{P}$.

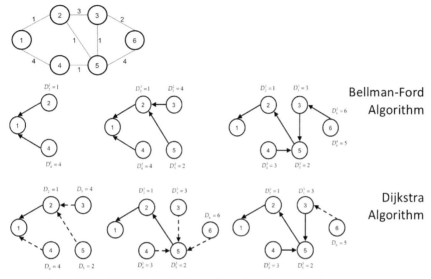

Figure 2.12 Illustration of bellman-form algorithm and dijkstra algorithm.

Proposition (Dijkstra Algorithm):

The initial condition is set as $\mathcal{P} = \{1\}$, $D_1 = 0$, $D_j = d_{j1}, \forall j \neq 1$. Repeat the following procedure.

(a) (Find the next closest node) Find $i \notin \mathcal{P}$ by

$$D_i = \min_{j \notin \mathcal{P}} D_j \tag{2.8}$$

Set $\mathcal{P} \leftarrow \mathcal{P} \cup \{i\}$. If $\mathcal{P}$ contains all nodes, stop.

(b) (Update labels) $\forall j \notin \mathcal{P}$,

$$D_j \leftarrow \min \left[D_j, d_{ij} + D_i \right] \tag{2.9}$$

Figure 2.12 illustrates the ways that Bellman-Ford Algorithm and Dijkstra Algorithm work by using a simple example. General applications of above algorithms can take advantage of the concept of *minimum spanning tree*. A *spanning tree* of a simple graph $\mathcal{G} = (V, E)$ is a sub-graph $\mathcal{T} = (V, E')$, which is a tree and has the same set of vertices as $\mathcal{G}$. If $w(e)$ is the *weight* of the edge e, a *minimum spanning tree* for $\mathcal{G}$ is a spanning tree such that $w(\mathcal{T}) \leq w(\mathcal{T}'), \forall \mathcal{T}'$.

▶ **Exercise (Touring in Florida):** Using the metric in Figure 2.1, please find the solution of touring in Florida.

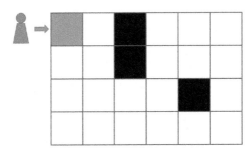

Figure 2.13   Robot path finding, to go through all 20 white tiles but must start and end on the green tile, while 3 black tiles are prohibited.

▶ **Exercise:** As the illustration of Figure 2.13, please plan the optimal path for the robot to clean the 20 white tiles. The robot must start and end on the green tile, and 3 black tiles are prohibited to access. The optimality is determined by the minimal number of movements (from one tile to its immediate 4 neighboring tiles being considered as one movement). What is the minimal number of movements in your optimal path(s)?

Since the involvement of cost into search algorithms, the complexity is hard to determine. Then, the uniform cost search can be considered as the worst case of complexity.

2.3.3 Depth-first Search

Different from breadth-first search, *depth-first search* always expands the deepest node in the current frontier of the search tree. Depth-first search proceeds immediately to the deepest level of the search tree (i.e. the node of no successors), then backs up to the next deepest node that still has unexplored successors. LIFO queue is most suitable for such search, as the most recent generated node is selected for expansion.

The time complexity of depth-first graph search is bounded by the size of the state space and appears no advantage over breadth-first search. However, depth-first search enjoys space complexity as only needs to store a single path from the root to a leaf node. For a state space of branching factor b and maximum depth d, the required storage space is only $O(bd)$ nodes.

A variant of depth-first search is called *backtracking search*, in which only one successor is generated at a time and each partially expanded node remembers corresponding successor to generate next. In this way, only $O(d)$ memory spaces is required.

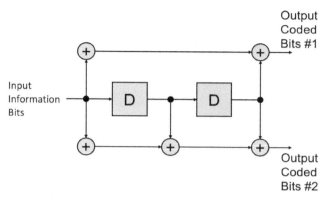

Figure 2.14 Rate 1/2 convolutional code encoder.

▶ **Exercise (Convolutional Codes):** Back to early days in developing statistical communication theory, *convolutional codes* have been known an effective class of forward error correcting codes to protect transmitted information bits. However, until 1968, the optimal decoding algorithm was first developed by A. Viterbi by so-called *Viterbi algorithm* that is widely used in digital communication system design such as *maximum likelihood sequence estimation, trellis coded modulations*, demodulation with channel memory, and *multiuser detection*. The Viterbi algorithm is actually a kind of backtracking search. Please consider the following convolutional encoder of rate $1/2$ (one input information bit generating two output coded bits). The encoder (thus decoder) usually starts from 00 and ends 00.

(a) Please find the state-transition diagram of this code.
(b) Please draw the possible trellis diagram (i.e. state transition diagram along with a sequence of input bits) for such code.
(c) Please explain how Viterbi algorithm based on likelihood (some combinations of bits are more likely than others according to state transition) works and reduces complexity of decoding.
(d) Please write Viterbi algorithm decoding for such code as a program. Please decode the sequence of received bits, 1110100011.

■ **Computer Exercise:** A robot is placed to walk through a maze as Figure 2.15 by starting from the exit in blue and getting out from the exit in red. The robot walks from one square to an immediate neighboring square (i.e. moving up, down, right, left). The black squares mean blocking (i.e. prohibited for a robot to get in, say a wall). The white squares mean the robot

Figure 2.15   Maze walking.

being allowed to move. Each time duration, a robot can sense its alternative movements (maybe more than one) and select its permissible movement in a random manner.

(a) Please develop and algorithm to walk the maze.
(b) Then, repeat the reverse direction (enter from red exit and leave the blue exit). What is the average steps for these two walks on this maze?
(c) Suppose the sensing (black square or not) of a robot has probability of error e. Up to now, we assume $e = 0$. If $0 < e \ll 1$, while the robot mistaking a black square to be white will waste one unit of time duration. Please repeat (b) for $e = 0.1$. Hint: Please consider the modifications of your algorithm in (a) and (b).

2.4 Informed Search

In many cases, we can take advantage of problem-specific knowledge beyond the definition of the problem itself, to find solutions in a more efficient manner, which is known as the *informed search* strategy. An immediate but general approach is called *best-first search* on top of general graph-search or tree-search, in which a node is selected for expansion based on an *evaluation function*, $\phi(n)$. Such evaluation function is constructed to estimate the cost, and the node of the lowest cost evaluation is expanded first. The implementation of best-first graph search is the same as Figure 2.9, except using ϕ instead of the order of priority queue.

Obviously, the selection of ϕ influences the search strategy. To facilitate the a best-first search algorithm, a *heuristic function*, $h(\cdot)$ is usually established to incorporate additional knowledge of the problem into the search algorithm. For example, in a shortest path problem, $h(n)$ can be the estimated cost of the cheapest path from the state at node n to the goal state. *Greedy best-first search* evaluates nodes by just using the heuristic function $h(\cdot)$, to expand the nodes closest to the goal state and lead to a quick solution. An example is to replace the distance metric in Figure 2.1 by the straight-line distance.

The most widely applied form of best-first search might be *A* search*, say for robot motion planning. The evaluation of nodes is facilitated by combining the cost to reach the node (i.e. $g(n)$) and the cost to get from the node to the goal (i.e. $h(n)$).

$$\phi(n) = g(n) + h(n) \tag{2.10}$$

Since $g(n)$ represents the path cost from the start node to node n, and $h(n)$ is the estimated cost of the cheapest path from node n to the goal, $\phi(n)$ therefore stands for the estimated cost of the cheapest solution through node n. Again, A* search is the same as uniform-cost search except using $g + h$ instead of g. A* search is not only reasonable but also complete and optimal, provided that the heuristic function satisfies some conditions.

- $h(\cdot)$ is an *admissible heuristic*, which implies that never overestimates the cost to reach the goal. Since $g(n)$ is the actual cost to reach node n along the current path, and $\phi(n) = g(n) + h(n)$, an immediate consequence suggests that $\phi(n)$ never overestimates the true cost of a solution. In other words, admissible heuristics are optimistic by nature.
- A slightly stronger condition called *consistency* (or sometimes *monotonicity*) is required only for applications of A* to graph search. A heuristic $h(n)$ is consistent if, for every node n and every successor n' of node n generated by any action a, the estimated cost of reaching the goal from node n is no greater than the step cost of getting to node n' plus the estimated cost of reaching the goal from node n', which implies a general triangle inequality:

$$h(n) \leq c(n, a, n') + h(n') \tag{2.11}$$

For an admissible heuristic, the inequality makes perfect sense that, if there is a route from node n to goal G_n via n' that costs less than $h(n)$, it would violate the property that $h(n)$ is a lower bound on the cost to reach G_n.

It is straightforward to show that every consistent heuristic is also admissible. Consistency is therefore a stricter requirement than admissibility, but almost all admissible heuristics are also consistent in practice.

Proposition (Optimality of A*):

A* has the following properties:

(a) The tree-search version of A* is optimal if $h(\cdot)$ is admissible.
(b) The graph-search version of A* is optimal if $h(\cdot)$ is consistent.

Proof: We prove the part (b) of the Proposition.

We first show that if $h(n)$ is consistent, then the values of $\phi(n)$ along any path are nondecreasing.The proof immediately follows the definition of consistency.

The next step is to prove that whenever A* selects a node n for expansion, the optimal path to that node has been found.

Therefore, we conclude that the sequence of nodes expanded by A* in graph-search is in nondecreasing order of $\phi(n)$. Hence, the first goal node selected for expansion must be an optimal solution because ϕ is the true cost for goal nodes (which have $h = 0$) and all later goal nodes will be at least as expensive.

Remark: For algorithms that extend search paths from the root and use the same heuristic information, A* is *optimally efficient* for any given consistent heuristic.

▶ **Exercise:** For each of the following statements, please show it if you consider true, or supply a counterexample if you consider false.

(a) Breadth-first search is a special case of uniform search.
(b) Depth-first search is a special case of best-first tree search.
(c) Uniform-cost search is a special case of A* search.

Memory can be a serious constraint in heuristic search. To reduce the required memory of A*, *iterative-deepening A** (IDA*) adapts iterative deepening to the heuristic search context, where *iterative deepening (depth-first) search* is a graph search strategy in which a depth-limited version of depth-first search is run repeatedly with increasing depth limits until the goal is found. The main difference between IDA* and standard iterative deepening is that the cutoff used is the ϕ-cost rather than the depth itself; at each iteration, the cutoff value is the smallest ϕ-cost of any node that exceeded the cutoff on the previous iteration. IDA* is practical for many problems with unit step costs and avoids the substantial overhead associated with keeping a sorted queue of nodes. Unfortunately, it suffers from the same difficulties with real-valued costs as does the iterative version of uniform-cost search. Two other memory-bounded algorithms are of interest:

- *Recursive best-first search* (RBFS) is a simple recursive algorithm that attempts to mimic the operation of standard best-first search, but using only linear space. Similar to recursive depth-first search but continuing indefinitely down the current path, ϕ-limit variable can be introduced to keep tracking ϕ-value of the best alternative path. Like A* tree search, RBFS is an optimal algorithm if the heuristic function $h(\cdot)$ is admissible. Its space complexity is linear in the depth of the deepest optimal solution, but its time complexity is usually difficult to characterize.
- It is also interesting to use up all memory to develop two algorithms, *memory-bounded A** (MA*) and *simplified MA** (SMA*). SMA* proceeds just like A*, expanding the best leaf until memory is full (i.e. it cannot add a new node to the search tree without dropping an old one). SMA* then drops the worst leaf node, the one with the highest ϕ-value. SMA* is complete, if there is any reachable solution; that is, if the depth of the shallowest goal node, is less than the memory size.

Remark: To enable an agent searching better by learning, *meta-level state space* that captures computational status can be introduced to substitute *object-level space* in the original search problem. A *meta-level learning* algorithm can learn from these experiences to avoid exploring unpromising subtrees to speed up search.

Remark: Heuristic functions obviously play a critical role in general search problems. To more efficiently develop the desirable solution, a *relaxed problem* that posts few restrictions on actions can be created. Because the relaxed problem actually amends edges to the original state space, any optimal solution in the original problem is, by definition, also a solution in the relaxed problem. However, the relaxed problem may have better solutions if the amended edges create short cuts. Consequently, the cost of an optimal solution to a relaxed problem is an admissible heuristic for the original problem. Furthermore, because the derived heuristic is an exact cost for the relaxed problem, it must obey the triangle inequality and is therefore consistent.

▶ **Exercises [1]:** n vehicles occupy squares $(1, 1)$ through $(n, 1)$ (i.e., the bottom row) of an $n \times n$ grid. The vehicles must be moved to the top row but in reverse order; so the vehicle i that starts in $(i, 1)$ must end up in $(n-i+1, n)$. On each time step, every one of the n vehicles can move one square up, down, left, or right, or stay put; but if a vehicle stays put, one other adjacent vehicle (but not more than one) can hop over it. Two vehicles cannot occupy the same square. Please answer the following questions:

(a) Calculate the size of the state space as a function of n.
(b) Calculate the branching factor as a function of n.
(c) Suppose that vehicle i is at (x_i, y_i); write a nontrivial admissible heuristic h_i for the number of moves it will require to get to its goal location $(n - i + 1, n)$, assuming no other vehicles are on the grid.
(d) Please explain whether the following heuristics are admissible for the problem of moving all n vehicles to their destinations: (i) $\sum_{i=1}^{n} h_i$; (ii) $\max\{h_1, \cdots, h_n\}$; (iii) $\min\{h_1, \cdots, h_n\}$.

■ **Computer Exercise:** Consider a crossword puzzle as the Figure 2.16 by using the 24 candidate words on the right-hand side of the figure. Each of the vertical labels, 1, 2, 3, and the horizontal labels, a, b, c, must be a word from the candidates. Each candidate word can be used only once. Please develop the algorithm to find all possible solutions, and identify the complexity of your algorithm. Is your algorithm computationally efficient based on number of steps in search (with calculations)? You must specify details such as knowledge representation, pruning methods, etc.

■ **Computer Exercise:** LeBran finds a part-time job to pack products into a box of size $12.5 \times 7.5 \times 5.5$ (in inch). If he packs a product A of size $4 \times 3 \times 2$ into box, he can earn \$3. If he packs a product B of size $3 \times 2 \times 2$, he can earn \$1.5. If he packs a product C of ball with radius 1, he can earn \$1. Each box must contain at least one product A and one product B.

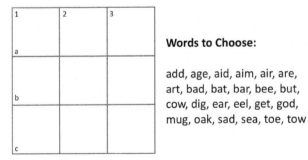

Figure 2.16 Crossword puzzle.

(a) How can LeBran earn most money?
(b) If the empty space must fill in soft medium to avoid instability, and each cubic inch of soft medium costs LeBran $0.25. How can LeBran earn most money?

2.5 Optimization

A search algorithm may output a number of solutions[1] and then we have to adopt some performance measure to identify the optimal solution(s). Therefore, *optimization* emerges as a critical problem in all kinds of AI, robotics, and machine learning problems.

An optimization problem, in general, has the following mathematical form:

$$maximize \quad f_0(x) \qquad (2.12)$$

$$subject\ to\ f_i(x) \le \beta_i \qquad (2.13)$$

where *minimize* is another (and usually mathematically equivalent) form of optimization problems.

Let the vector $\mathbf{x} = (x_1, \cdots, x_n)$ represent the vector of optimization variables in an optimization problem. $f_0 : \mathsf{R}^n \to \mathsf{R}$ denotes the objective function, and $f_i : \mathsf{R}^n \to \mathsf{R}, i = 1, \cdots, m$ denote the constraints functions, where $\beta_1, \cdots, \beta_m$ are the bound or limit of each constraint. $\mathbf{x}^*$ is called *optimal solution* of the optimization problem in (2.12) and (2.13), if $\mathbf{x}^*$ gives the largest objective value among all possibilities satisfying the constraints. That is,

$$f_1(z) \le \beta_1, \cdots, f_m(z) \le \beta_m, f_0(z) \le f_0(\mathbf{x}^*), \forall z \qquad (2.14)$$

When the optimization problem is *minimize* in (2.12), $f_0(z) \ge f_0(\mathbf{x}^*)$ in (2.14). Several classes of optimization problems are of particular interest and introduced in the following.

2.5.1 Linear Programming

A function $\psi(\cdot)$ is called *linear* by satisfying $\forall x, y \in \mathsf{R}^n, \forall a, b \in \mathsf{R}$

$$\psi(ax + by) = a\psi(x) + b\psi(y) \qquad (2.15)$$

---

[1]It is possible that no solution is available.

The optimization of (2.12) and (2.13) is known as a *linear program* if the objective and constraint functions $f_0, f_1, \cdots, f_m$ are linear. An important class of optimization problem is *linear programming* as

$$maximize \qquad \mathbf{c}^T \mathbf{x} \tag{2.16}$$

$$subject\ to\ \ \mathbf{a_i}^T \mathbf{x} \leq \beta_i, i = 1, \cdots, m \tag{2.17}$$

where $\mathbf{c}, \mathbf{a_i} \in \mathsf{R}^n, i = 1, \cdots, m$ and $\beta_1, \cdots, \beta_m \in \mathsf{R}$.

   Although there does not exist simple analytical solution of a linear program, there exists a variety of effective methods to solve, such as Dantzig simplex method, of complexity $O(n^2 m)$ by assuming $m \geq n$.

Example (Chebyshev Approximation): Let us consider the *Chebyshev approximation* problem as

$$\min \max_{i=1,\cdots,k} | \mathbf{a_i}^T \mathbf{x} - \beta_i | \tag{2.18}$$

where $\mathbf{x}$ is the variable, and $\mathbf{a_1}, \cdots, \mathbf{a_k} \in \mathsf{R}^n, \beta_1, \cdots, \beta_k \in \mathsf{R}$ are parameters. It can be solved by solving the linear program

$$minimize \qquad \tau$$
$$subject\ to\ \ \ \mathbf{a_i}^T \mathbf{x} - \tau \leq \beta_i, i = 1, \cdots, k$$
$$-\mathbf{a_i}^T \mathbf{x} - \tau \leq -\beta_i, i = 1, \cdots, k$$

where $\mathbf{x} \in \mathsf{R}^n, \tau \in \mathsf{R}$.

▶ **Exercise:** Please use the simplex method to compute the following linear programming

$$Minimize \qquad x - 3y$$
$$Subject\ to\ -x + 2y \leq 6$$
$$x + y \leq 5$$
$$x, y \geq 0$$

2.5.2 Nonlinear Programming

If the optimization problem is not linear, it is a *nonlinear program*. The nonlinear optimization has to deal with complex functional and usually difficult to find the global solution(s). Instead, local optimization algorithms have been widely developed. In other words, there is no general means to solve nonlinear programming.

2.5.3 Convex Optimization

A special class of non-linear optimization problem is the *least-squares* (LS) problem, which generally has no constraint and the objective function is the sum of squares in the form $\mathbf{a_i}^T\mathbf{x} - \beta_i$:

$$Minimize\ f_0(\mathbf{x}) = \|\mathbf{A}\mathbf{x} - \beta\|^2 = \sum_{i=1}^{k}(\mathbf{a_i}^T\mathbf{x} - \beta_i)^2 \tag{2.19}$$

where $\mathbf{a_i}^T$ are the rows of $\mathbf{A} \in \mathrm{R}^{k \times n}$, $k \geq n$, β is the vector of β_i's, and $\mathbf{x} \in \mathrm{R}^n$ is the variable vector in the optimization.

This LS problem is widely seen in signal processing and machine learning, and can be solved by a set of linear equations. By treating

$$(\mathbf{A}^T\mathbf{A})\mathbf{x} = \mathbf{A}^T\beta, \tag{2.20}$$

we obtain

$$\mathbf{x} = (\mathbf{A}^T\mathbf{A})^{-1}\mathbf{A}^T\beta. \tag{2.21}$$

The LS problem has a useful varient by introducing weights (i.e. $w_1, \cdots, w_k$) into optimization as the *weighted least-squares*

$$f_0(\mathbf{x}) = \sum_{i=1}^{k} w_i(\mathbf{a_i}^T\mathbf{x} - \beta_i)^2 \tag{2.22}$$

Another useful technique to solve LS problems is *regularization* by introducing an extra term to control the convergence.

$$f_0(\mathbf{x}) = \sum_{i=1}^{k}(\mathbf{a_i}^T\mathbf{x} - \beta_i)^2 + \lambda \sum_{i=1}^{n} x_i^2 \tag{2.23}$$

Actually, (2.23) can be re-written into a more general form in Banach space.

$$f_0(\mathbf{x}) = \sum_{i=1}^{k} \|\mathbf{a_i}^T\mathbf{x} - \beta_i\|_2 + \lambda\|\mathbf{x}\|_q \tag{2.24}$$

where the regularization term is in the q-th norm but we often consider the cases for $q = 0, 1, 2$. It is extremely useful in the *regression* of *statistical learning*.

Among wide non-linear optimization problems, a class of problems turns out of high interest, *convex optimization* that has the following form:

$$Minimize \qquad f_0(x) \qquad\qquad (2.25)$$

$$Subject\ to\ \ f_i(x) \le \beta_i, i = 1, \cdots, m \qquad\qquad (2.26)$$

where the functions $f_0, f_1, \cdots, f_m : \mathsf{R}^n \to \mathsf{R}$ are convex by satisfying

$$f_i(ax + by) \le a f_i(x) + b f_i(y), \forall x, y \in \mathsf{R}^n, \forall a, b \ge 0, a + b = 1 \quad (2.27)$$

The LS problems and linear programming problems are special cases of convex optimization. Although convex optimization is hard to analytically solve, computationally effective algorithms are available. Consequently, convex optimization attracts tremendous attention in recent technology development of systems engineering, machine learning, and data analytics.

▶ **Exercise:** Consider the following problem:

$$Minimize\ \ (x - 4)^2 + (y - 6)^2 \qquad\qquad (2.28)$$

$$Subject\ to \qquad y \ge x^2 \qquad\qquad (2.29)$$

$$y \le 4 \qquad\qquad (2.30)$$

(a) What is the necessary condition for optimality?
(b) Is the point $(2, 4)$ satisfying (a)? Is this the optimal point? Please explain.

Further Reading: [1] provides detailed and holistic search algorithms of artificial intelligence. [2] supplies comprehensive knowledge in algorithms on graphs. For readers intending to know more theoretical treatments on dynamic programming, [3] supplies in-depth knowledge of aspects from decision and control. For in-depth understanding of convex optimization, [4] is excellent to read.

References

[1] Stuart Russell, Peter Norvig, *Artificial Intelligence: A Modern Approach*, 3rd edition, Prentice-Hall, 2010.
[2] J. Kleinberg, E. Tardos, *Algorithm Design*, 2nd. Edition, Pearson Education, 2011.
[3] D. P. Bertsekas, *Dynamic Programming*, Prentice-Hall, 1987.
[4] S. Boyd, L. Vandenberghe, *Convex Optimization*, Cambridge University Press, 2004.

3

Machine Learning Basics

Machine learning has been introduced as a marvelous technique for realizing artificial intelligence since the late 1950's. Machine learning algorithms can learn from and make decisions on given training data without being explicitly programmed. In 1959, the term *machine learning* (ML) was first proposed by Arthur Samuel, which grants computer systems the capability of learning with large amounts of previous tasks and data, as well as of self-optimizing computer algorithms.

Example: The power of mathematical statistics can be illustrated by this study in political science [1]. Researchers analyze the statistics of voting in various countries based on public available information by creating 2-D histograms of the number of units, as shown in the following figure with x-axis representing the percentage of voter turnout and y-axis as the percentage of votes for the winner. We can actually learn a lot from this simple statistical presentation. The red circles as a separate cluster suggest near 100% turnout ratio and near 100% votes to the winner. What can you learn from such an election fingerprint? Another interesting point is for Canada with two clusters, due to French speaking Canada and English speaking Canada. We therefore can infer or learn from statistical data.

Based on the capability of self-learning, machine learning is beneficial for classifying/regressing, predicting, clustering and decision making. Machine learning has the following three basic elements [2]:

- *Model*: Mathematical models are abstracted from training data and expert knowledge in order to statistically describe the characteristics or objective laws of the given data set. Assisted by these trained models, machine learning can then be used for classification, prediction and decision making.

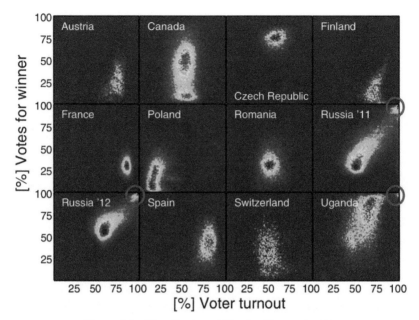

Figure 3.1   Visualization of Election Fingerprint [1].

- *Strategy*: The criteria for training mathematical models are called strategies. How to select an appropriate strategy is closely associated with training data. Empirical risk minimization (ERM) and structural risk minimization (SRM) are two fundamental strategic issues, where the latter can beneficially avoid the *over-fitting* phenomenon when the sample size is small.
- *Algorithm*: Algorithms are constructed to solve unknown parameters based on the determined model and selected strategy, which can be viewed as an optimization process. A good algorithm can not only yield a globally optimal solution, but also has low computational complexity and storage complexity.

Statistical learning theory was introduced in the late 1960's and is considered as a branch of mathematical statistical analysis treating the problem of function estimation from a given collection of data. Particularly since the invention of widely applied *support vector machines* (SVMs) in the mid-1990's, statistical learning theory has been shown to be useful to develop new learning algorithms.

3.1 Supervised Learning

If machine learning proceeds as there is a teacher to supply feedback for the model or the algorithm, it is called *supervised learning*. In other words, a training dataset is typically available for supervised learning.

3.1.1 Regression

Regression analysis can be viewed as a kind of statistical process method for estimating the relationships among variables. Relying on modeling the function relationship between a dependent variable (objective) and one or more independent variables (predictors), regression is a powerful statistical tool for predicting and forecasting a continuous-valued objective given a set of predictors.

In the regression analysis, there are three variables, i.e.

- *Independent variables* (predictors): X
- *Dependent variable* (objective): Y
- *Other unknown parameters* that affect the estimate value of the dependent variable: ε

The regression function f models the function relationship between X and Y perturbed by ε, which can be given by: $Y = f(X, \varepsilon)$. Usually, we characterize the variation of the predictors X around the objective Y in terms of a specific regression function with a probability distribution. Moreover, the approximation is often modeled as $E = [Y \mid X] = f(X, \varepsilon)$. When conducting regression analysis, first of all it needs to determine the form of regression function f, which relies on both the common knowledge about the relationship between dependent variable and independent variables as well as on the principle of convenient computing. Based on the form of regression function, the methods of regression analysis can be differentiated, such as ordinary linear regression, logistic regression, polynomial regression, etc.

In the linear regression, the dependent variable is a linear combination of the independent variables or unknown parameters. Suppose N random training samples with M independent variables, i.e. $\{y_n, x_{n1}, x_{n2}, \ldots, x_{nM}\}, n = 1, 2, \ldots, N$, the linear regression function can be formulated as:

$$y_n = \beta_0 + \beta_1 x_{n1} + \beta_2 x_{n2} + \cdots + \beta_M x_{nM} + e_n, \qquad (3.1)$$

where β_0 is termed as regression intercept, while e_n is the error term and $n = 1, 2, \ldots, N$. Hence, Equation (3.1) can be rewritten in the form of matrix as $y = X\beta + e$, where $y = [y_1, y_2, \ldots, y_N]^T$ is an observation vector of the dependent variable and $e = [e_1, e_2, \ldots, e_N]^T$, while $\beta = [\beta_0, \beta_2, \ldots, \beta_M]^T$ and X represents the observation matrix of independent variables, i.e.

$$X = \begin{bmatrix} 1 & x_{11} & \cdots & x_{1M} \\ 1 & x_{21} & \cdots & x_{2M} \\ \vdots & \vdots & \ddots & \vdots \\ 1 & x_{N1} & \cdots & x_{NM} \end{bmatrix}.$$

Linear regression analysis aims for estimating unknown parameter $\widehat{\varepsilon}$ relying on the least squares (LS) criterion. The solution can be expressed as:

$$\widehat{\beta} = (X^T X)^{-1} X^T y. \tag{3.2}$$

Example: A common simplified scenario of interest is to use input data

$$(x_1, y_1), \ldots, (x_p, y_p)$$

to obtain the linear predictor

$$\hat{Y} = \hat{b}_0 + \sum_{j=1}^{p} \hat{b}_j X_j \tag{3.3}$$

where $\hat{b}_0$ is the *bias*, which can be included in the $X = (X_1, \ldots, X_p)$. Then, defining $b = (b_1, \ldots, b_p)$,

$$\hat{Y} = X^T \hat{b} \tag{3.4}$$

Applying the *least squared error* to measure, we can identify the weighting coefficient vector b by minimizing the *residual sum of squares*

$$RSS(\hat{b}) = \sum_{i=1}^{p} (y_i - x_i^T b)^2 \tag{3.5}$$

$\hat{b}$ is a quadradic function and hence the minimum always exists but may not be unique.

$$RSS(\hat{b}) = (y - Xb)^T (y - Xb) \tag{3.6}$$

Differentiating with respect to b, we get the normal equations as (3.2)

$$\hat{b} = (X^T X)^{-1} X^T y \tag{3.7}$$

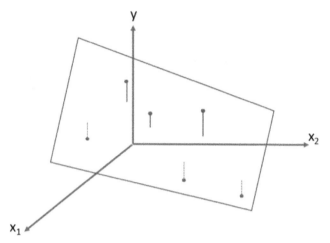

Figure 3.2 Visualization of Multiple Regression Hyperplane for $M = 2$ (i.e. 2-D hyperplane shown in red) and Regression Points with Errors (i.e. solid line indicating on top of the regression plane and dot line indicating under the regression plane).

This example of simple regression, $M = 1$, has been widely used to infer sensed or observed data in robotics. The intuitive visualization of *multiple regression* ($M \geq 2$) is illustrated in Figure 3.2.

■ **Programming Exercise:** Please collect the stock price data about Citi Bank, Morgan Stanley, INTEL, Amazon, Boeing, Johnson & Johnson, PG&E, Exxon Mobile, and gold price in 2018. We try to use these 8 stocks to infer gold price by regression.

(a) Please use each of the 8 stocks to predict gold price on the first trading date in 2019, to find which stock supplies best accuracy.

(b) Please use all 8 stocks for multiple regression to predict the gold price on the first trading day in 2019. Compare with the result in (a). Explain your result.

(c) Which one(s) among the 8 stocks, can be used to supply better prediction based on regression methods? *Hint: Are all 8 stocks equally useful? Equally conditionally useful?*

(d) Suppose the predictor y_n for the nth day is composed as

$$y_n = w_1 y_{n-1} + w_2 y_{n-2} + \cdots + w_l y_{n-l}$$

which actually represents a kind of digital filtering in estimation. Please repeat problem $3(a)$ by identifying optimal depth of l and subsequently the weighting coefficients $w_1, \ldots, w_l$.

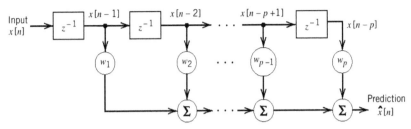

Figure 3.3   Linear Prediction Filter of Order p.

▶ **Exercise (Wiener Filter):** Suppose we intend to develop a hardware structure for prediction, which is a common problem in (digital) signal processing to implement by a *finite duration impulse response* filter as shown in Figure 3.3. Such a linear prediction FIR has the following three functional blocks:

- p unit-delay elements
- multipliers with weighting coefficients $w_1, w_2, \ldots, w_p$
- adders to sum over the delayed inputs to generate the output prediction $\hat{x}[n]$

The input signal is assumed from the sample function of a wide-sense stationary process $X(t)$ of zero mean and autocorrelation function $R_X(\tau)$. Please design the filter by identifying *Wiener-Hopf* equation to specify the coefficients, which is equivalent to develop the following Theorem.

Theorem (Wiener-Hopf Equation): Let
$\boldsymbol{w_o} = [w_1, \ldots, w_p]$ denote $p \times 1$ coefficient vector of optimum filter;
$\boldsymbol{r_X} = [R_x[1], \ldots, R_X[p]]^T$ denote $p \times 1$ autocorrelation vector;

$$\boldsymbol{R_x} = \begin{bmatrix} R_X[0] & R_X[1] & \cdots & R_X[p-1] \\ R_X[1] & R_X[0] & \cdots & R_X[p-2] \\ \vdots & \vdots & \ddots & \vdots \\ R_X[p-1] & R_X[p-2] & \cdots & R_X[0] \end{bmatrix}$$

denote $p \times p$ autocorrelation matrix. Then,

$$\boldsymbol{w_o} = \boldsymbol{R_X}^{-1}\boldsymbol{r_X} \tag{3.8}$$

Remark: The primary difference between above exercise and immediately earlier programming exercise lies in the modeling **x** (i.e. input data or training data) as data from a stationary random process with known statistics. Under this sense, the resulting *Wiener filter* is optimal, which forms the

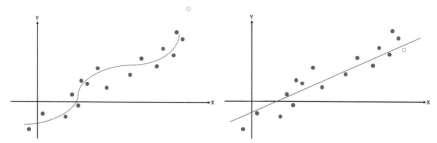

Figure 3.4   Formation of Overfitting in Machine Learning. (a) High order regression curve well fits the training data with minimum squared error, and thus predicts/infers the blue circle point. (b) Given the same training dataset, linear regression can deliver a good prediction/inference as red circle point that is actually close to the true value, but quite different from high order regression.

foundation in statistical signal processing with applications to control and communication systems.

In regression or general machine learning methodology, once there exists a training set, we have to be careful about one phenomenon known as *overfitting*, which can be explained by Figure 3.4(a) and 3.4(b). In short, more complex methods to better fit the training data do not necessarily deliver accurate inference (i.e. prediction in the figure).

▶ **Exercise:** Antonio collects a dataset of 4 points, and each has an input value and output value from an unknown system. These 4 points are

$$(0.98, 0.21), (2.09, 1.42), (2.83, 2.12), (3.84, 4.15).$$

Based on these 4 points, we intend to find the best prediction for input value 5.0 by using a polynomial function.

(a) What is the best prediction in the form of $y = c_0 + c_1 x + c_2 x^2 + \cdots$ for input value 5.0 based on the training data?
(b) If the true output value for the 5th point is 3.96, how do you explain your inaccuracy in (a)?

The next question for simple regression is to measure the *goodness of fit* for the regression. The *coefficient of determination*, r^2, can measure how well the linear approximation produced by the least-squares regression line (or hyperplane) actually fits the observed data. We define the *sum of squares total* (SST) as

$$SST = \sum_{i=1}^{n} (y - \bar{y})^2 \tag{3.9}$$

and the *sum of squares regression* (SSR) as

$$SSR = \sum_{i=1}^{n}(\hat{y} - \bar{y})^2 \tag{3.10}$$

Then,

$$r^2 = \frac{SSR}{SST} \tag{3.11}$$

serves our purpose.

More than a century ago in 1914, A. Einstein (yes, the genius to create the theory of relativity) first raised the statistical analysis on time series of observations. A time series is a set of observations x_t, t countable, suggesting a discrete-time series. We intend to develop techniques to analyze and to draw inferences from x_t. Time series analysis, as one of the main subjects in statistical signal processing has been widely used in tracking, navigation, acoustic and image processing, remote sensing, information retrieval, and finance and economy in particular. In other words, time series deal with the data of time index.

Example: The daily closing prices of a stock in 2018 form a time series.

Definition: A time series model for the observed data $\{x_t\}$ is a specification of the joint distributions of a sequence of random variables $\{X_t\}$ of which $\{x_t\}$ can be treated as a realization of such random process.

Definition: MA(1) The well-known *moving average* (MA) time series can be represented by

$$X_t = Z_t + cZ_{t-1} \tag{3.12}$$

Considering the sampling times $t_n = nT_s$ (or, any series of embedded timings) as digital signal processing, we may re-write above equation into

$$x[n] = Z[n] + xZ[n-1] \tag{3.13}$$

Definition AR(1)**:** Another widely known model for time series is *auto-regressive* (AR) as

$$X_t = cX_{t-1} + W_t \tag{3.14}$$

where $W_t \sim G(0, \sigma^2)$ is a Gaussian distribution with zero mean and variance σ^2 at time t. In digital signal processing format, we can re-write as

$$X[n] = cX[n-1] + W[n] \tag{3.15}$$

Remark: Please note the infinite time response suggested from the equation for AR models. Please also note that $\{X_t\}$ is time varying in general. However, to the ease of mathematical manipulations, a stationary process (and thus a sequence of random variables) may be assumed.

▶ **Exercise:** The time series for stock prices of BayTech Inc. in past 5 trading days is $x_1 = 23.35, x_2 = 24.42, x_3 = 25.31, x_4 = 24.96, x_5 = 24.37$, and use regression method to predict the stock price next day (i.e. x_6).

Another type of regression, *logistic regression*, considers the categorical variables, for example, binary. In order to facilitate our analysis, in the following we consider the case of a binary dependent variable, for example. The goal of the binary logistic regression is to model the probability of the dependent variable being the value 0 or 1 given training samples. To elaborate a little further, let the binary dependent variable y depends on M independent variables $\boldsymbol{x} = [x_1, x_2, \ldots, x_M]$. The conditional distribution of y under the condition of $\boldsymbol{x}$ is a Bernoulli distribution. Hence, the probability of $\Pr(y = 1 \mid \boldsymbol{x})$ can be formulated in the form of a standard logistic function,[1] also called sigmoid function:

$$P \triangleq \Pr(y = 1 \mid \boldsymbol{x}) = \frac{1}{1 + e^{-g(\boldsymbol{x})}}, \tag{3.16}$$

where $g(\boldsymbol{x}) = w_0 + w_1 x_1 + w_2 x_2 + \cdots + w_M x_M$ and $\boldsymbol{w} = [w_0, w_1, \ldots, w_M]$ represents the regression coefficient vector. Similarly,

$$\Pr(y = 0 \mid \boldsymbol{x}) = 1 - P = \frac{1}{1 + e^{g(\boldsymbol{x})}}. \tag{3.17}$$

Relying on aforementioned definitions, we have $g(\boldsymbol{x}) = \ln(\frac{P}{1-P})$. Hence, for a given dependent variable, the probability of its value being y_n can be given by $P(y_n) = P^{y_n}(1 - P)^{1-y_n}$. Given a set of training samples $\{y_n, x_{n1}, x_{n2}, \ldots, x_{nM}\}, n = 1, 2, \ldots, N$, we are capable of estimating the regression coefficient vector $\boldsymbol{w} = [w_0, w_1, \ldots, w_M]$ with the aid of the maximum likelihood estimation (MLE) method. Explicitly, logistic regression can be deemed as a special case of the generalized linear regression family.

---

[1]The logistic function is a common "S" shape function, which is the cumulative distribution function (CDF) of the logistic distribution.

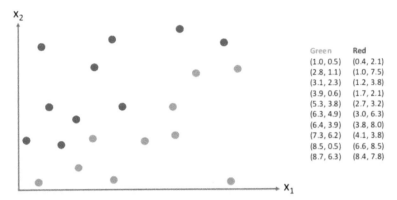

Figure 3.5   Clustering of dots.

■ **Exercise:** There are 10 green dots and 10 red dots as shown in Figure 3.5, with their coordinates. In order to separate these dots to green group and red group under the criterion of least squares:

(a) Please identify a linear regression curve to separate these dots.
(b) Please consider a kernel for regression (in other words, logistic regression) to separate these dots. *Hint: A simple version of sigmoid function will serve the purpose* of logistic regression in this case.
(c) Another possible approach proceeds on rule-based *decision tree* that will be introduced in Chapter 9.

3.1.2 Bayesian Classification

The Bayes classifier is a family member of probabilistic classifiers relying on the Bayes' theorem by computing the posteriori probability distribution of the objective variable given a set of training samples. As a widely-used classification method, for example, The naive Bayes classifier can be trained efficiently conditioned on a simple but strong independence assumptions among features. Furthermore, training a naive Bayes model has a linear time complexity with the aid of MLE, which yields a closed-form expression.

Let vector $\boldsymbol{x} = [x_1, x_2, \ldots, x_M]$ represent M independent features for a total of K classes $\{y_1, y_2, \ldots, y_K\}$. For each of K possible class label y_k, we have the conditional probability of $p(y_k|x_1, \ldots, x_M)$. Relying on Bayes' theorem, we decompose the conditional probability in the form of:

$$p(y_k|x_1, \ldots, x_M) = \frac{p(y_k)p(x_1, \ldots, x_M|y_k)}{p(x_1, \ldots, x_M)}, \qquad (3.18)$$

where $p = (y_k|x_1, \ldots, x_M)$ is called posteriori probability, whilst $p(y_k)$ is the priori probability of y_k. Given that x_i is conditionally independent of x_j for $i \neq j$, we have:

$$p(y_k|x_1, \ldots, x_M) = \frac{p(y_k)}{p(x_1, \ldots, x_M)} \prod_{m=1}^{M} p(x_m|y_k), \qquad (3.19)$$

where $p(x_1, \ldots, x_M)$ only depends on M independent features which can be viewed as a constant.

Maximum a posteriori (MAP) is used as the decision making rule for the naive Bayes classifier. Given a feature vector $\overline{x} = (\overline{x}_1, \overline{x}_2, \ldots, \overline{x}_M)$, its label $\overline{y}$ can be determined by:

$$\overline{y} = \underset{y_k \in \{y_1, \ldots, y_K\}}{\arg \max} \; p(y_k) \prod_{m=1}^{M} p(\overline{x}_m|y_k) \qquad (3.20)$$

Despite its easy implementation and oversimplified assumptions, naive Bayes classifiers have witnessed their success in numerous complex real-world situations, such as outlier detection, spam filtering, etc. More concepts will be introduced in Chapter 4 about statistical decisions.

▶ **Exercise (Pattern Recognition):** One popular class of problems in Bayesian classification, including statistical decision in Chapter 4, is *pattern recognition*. An orthonormal basis, $\{\phi_n(\cdot)\}_1^N$, is usually selected to represent patterns of interest, where

$$\int \phi_i(\tau)\phi_j(\tau)d\tau = \delta_{ij} \qquad (3.21)$$

A pattern $s(\tau)$ can be represented (i.e. expanded) by such an orthonormal basis as

$$s(\tau) = \sum_{n=1}^{N} s_n \phi_n(\tau) \qquad (3.22)$$

such that we can more precisely evaluate the posteriori probability for the Bayes classifier. On the top of Figure 3.6, there are 6 reference numbers that are marked by 3×5 squares and are equally probable.

(a) For the red pattern, please develop the methodology and algorithm to recognize the number. Please note no vertical nor horizontal alignment for red pattern as reference numbers.

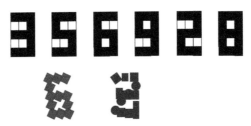

Figure 3.6   6 Reference Numbers on the Top and 2 Patterns to be Recognized.

    (b) For the purple pattern, please modify your approach in (a) to recognize the number.

A robot usually has to recognize some reference points but not from a perfect angle nor under a perfect condition. This exercise illustrates such a situation.

3.1.3 KNN

K nearest neighbors (KNN) is a non-parametric and instance-based learning method, which can be used for both classification and regression. Proposed by Cover and Hart in 1968, the KNN algorithm is one of the simplest of all machine learning algorithms. Relying on measuring the distance between the object and training samples in a feature space, a KNN algorithm determines the class or property value of the object. Specifically, in a classification scenario, an object is categorized into a specific class by a majority vote of its K nearest neighbors. If $K = 1$, the category of the object is the same with that of its nearest neighbor, and this case is termed as the one nearest neighbour classifier. By contrast, in a regression scenario, the output value of the object is calculated by the average of the value of its K nearest neighbors. Figure 3.7 shows the illustration of the unweighted KNN mechanism with $K = 4$.

    Suppose there are N training sample pairs, i.e. $\{(\boldsymbol{x}_1, y_1), (\boldsymbol{x}_2, y_2), \ldots, (\boldsymbol{x}_N, y_N)\}$, where y_n is the property value or class label of the sample $\boldsymbol{x}_n$, $n = 1, 2, \ldots, N$. Usually, we use the Euclidean distance or the Manhattan distance to calculate the similarity between the object $\overline{\boldsymbol{x}}$ and training samples. Let $\boldsymbol{x}_n = [x_{n1}, x_{n2}, \ldots, x_{nM}]$ contain M different features. Hence, the Euclidean distance between $\overline{\boldsymbol{x}}$ and $\boldsymbol{x}_n$ can be given by:

$$d_e = \sqrt{\sum_{m=1}^{M} (\overline{x}_m - x_{nm})^2}, \tag{3.23}$$

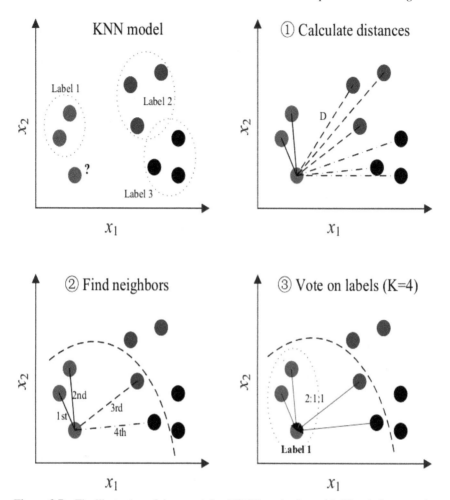

Figure 3.7   The illustration of the unweighted KNN mechanism with $K = 4$, for example.

while their Manhattan distance is calculated as:

$$d_m = \sqrt{\sum_{m=1}^{M} |\overline{x}_m - x_{nm}|}. \tag{3.24}$$

Relying on the similarity, the class label or property value of $\overline{x}$ can be voted or weightedly voted by its K nearest neighbors, i.e.

$$\overline{y} \leftarrow \text{VOTE} \{K \text{ nearest } (x_k, y_k)\}. \tag{3.25}$$

The performance of KNN algorithm largely depends on the value of K, and yet the best choice of K hinges upon the training samples. In general, a large K is conducive to resist the harmful interference from noisy data, while it blurs the class boundary between different categories. Fortunately, an appropriate value of K can be determined by a variety of heuristic techniques based on the characteristic of the training data set.

3.1.4 Support Vector Machine

SVM is another supervised learning model for classification and regression relying on constructing a hyperplane or a set of hyperplanes in a high-dimensional space. The best hyperplane is the one that results in the largest margin between classes. However, the training data set may often be not linearly separable in a finite dimensional space. To address this issue, SVM is capable of mapping the original finite dimensional space into a higher dimensional space, where the training data set can be more easily discriminated in that space.

Considering a linear binary SVM, for example, there are N training samples in the form of $\{(\boldsymbol{x}_1, y_1), (\boldsymbol{x}_2, y_2), \ldots, (\boldsymbol{x}_N, y_N)\}$, where $y_n = \pm 1$ indicates the class label of the point $\boldsymbol{x}_n$. SVM aims for searching for a hyperplane with the maximum margin against the training samples, which best discriminates the two classes of $\boldsymbol{x}_n$ with $y_n = 1$ and $y_n = -1$. Here, the maximum margin implies a maximum distance between the nearest point and the hyperplane. The hyperplane is represented by:

$$\boldsymbol{\omega}^T \boldsymbol{x} + b = 0. \tag{3.26}$$

Hence, we can quantify the margin of the training sample $(\boldsymbol{x}_n, y_n)$ as:

$$\gamma_n = y_n(\boldsymbol{\omega}^T \boldsymbol{x}_n + b). \tag{3.27}$$

Moreover, we assume a correct classification if $\boldsymbol{\omega}^T \boldsymbol{x}_n + b \geq 0$ when $y_n = 1$, while $\boldsymbol{\omega}^T \boldsymbol{x}_n + b \leq 0$ when $y_n = -1$. Because of $y_n(\boldsymbol{\omega}^T \boldsymbol{x}_n + b) \geq 0$, and hence a large margin means a superior correct classification. SVM tries to find an optimal hyperplane that maximizes the minimum margin between training samples and the hyperplane considered. Given a set of linearly separable training samples, after the operation of normalization, SVM can be formulated as the following optimization problem:

$$\max_{\boldsymbol{w},b} \min_{n=1,\ldots,N} y_n \left(\left(\frac{\boldsymbol{w}}{\|\boldsymbol{w}\|} \right)^T \boldsymbol{x}_n + \frac{b}{\|\boldsymbol{w}\|} \right)$$

$$\text{s.t. } y_n(\boldsymbol{w}^T \boldsymbol{x}_n + b) \geq \gamma, n = 1, 2, \ldots, N, \tag{3.28}$$

$$\|\boldsymbol{w}\| = 1,$$

where $\gamma = \min_{n=1,\ldots,N} y_n((\frac{\boldsymbol{w}}{\|\boldsymbol{w}\|})^T \boldsymbol{x}_n + \frac{b}{\|\boldsymbol{w}\|})$. Through some mathematical manipulation, problem (3.28) can be reduced to an optimization problem with a convex quadratic objective function and linear constraints, which can be given by:

$$\min_{\boldsymbol{w},b} \frac{1}{2}(\|\boldsymbol{w}\|)^2$$

$$\text{s.t. } y_n(\boldsymbol{w}^T \boldsymbol{x}_n + b) \geq 1, n = 1, 2, \ldots, N. \tag{3.29}$$

Relying on Lagrange duality, we can obtain the optimal $\boldsymbol{w}$ and b.

If the training samples are non-linearly separable, SVM is capable of mapping data to a high dimensional feature space with a high probability of being linearly separable. This may result in a non-linear classification or regression in the original space. Fortunately, kernel functions play a critical role in avoiding the "curse of dimensionality" in the above-mentioned dimensionality ascending procedure. There are a variety of alternative kernel functions, such as linear kernel function, polynomial kernel function, radial basis kernel function, neural network kernel function, etc. Furthermore, some regularization methods haven been conceived for in order to make SVM be less sensitive to outlier points.

3.2 Unsupervised Learning

In this section, we will highlight some typical unsupervised learning algorithms, i.e. K-means clustering, expectation-maximization (EM), principal component analysis (PCA) and independent component analysis (ICA).

3.2.1 K-Means Clustering

K-means clustering is a distance based clustering method that aims for dividing N unlabeled training samples into K different cohesive clusters, where each sample belongs to one cluster. K-means clustering measures the similarity between two samples in terms of their distance. K-means clustering is often comprised of two major steps, i.e. assigning each training sample to one of K clusters in terms of the closest distance between the sample and given cluster centroids, and updating each cluster centroid with the mean of the samples assigned to it. The whole algorithm is hence implemented by repeatedly carrying out above-mentioned two steps until the convergence is achieved.

To elaborate a little further, given a set of samples $\{\boldsymbol{x}_1, \boldsymbol{x}_2, \ldots, \boldsymbol{x}_N\}$, where $\boldsymbol{x}_n = [x_{n1}, x_{n2}, \ldots, x_{nM}]$ is a M-dimensional vector. Let $\mathbb{S} = \{s_1, s_2, \ldots, s_K\}$ represent the cluster set, and $\boldsymbol{\mu}_k$ is the mean of the samples in s_k. K-means clustering intends to find an optimal cluster segmentation, which follows:

$$\mathbb{S}^* = \underset{\{s_1, s_2, \ldots, s_K\}}{\arg\min} \sum_{k=1}^{K} \sum_{\boldsymbol{x} \in s_k} \|\boldsymbol{x} - \boldsymbol{\mu}_k\|^2. \tag{3.30}$$

However, Equation (3.30) is a non-deterministic polynomial acceptable problem. Fortunately, there are a range of efficient heuristic algorithms, which converge quickly fast to a local optimum.

As one of famous low-complexity iterative refinement algorithms for K-means clustering, Lloyd's algorithm often yields satisfactory performance after a small number of iterations. Specifically, given K initial cluster centroid $\boldsymbol{\mu}_k, k = 1, \ldots, K$, the Lloyd's algorithm achieves the final cluster segmentation result by alternating between the following two steps, i.e.

- Step 1: In iterative round r, assign each sample to a cluster. For $n = 1, 2 \ldots, N$ and $i, k = 1, 2 \ldots, K$, if we have:

$$s_i^{(r)} = \{\boldsymbol{x}_n : \|\boldsymbol{x}_n - \boldsymbol{\mu}_i^{(r)}\|^2 \le \|\boldsymbol{x}_n - \boldsymbol{\mu}_k^{(r)}\|^2, \forall k\}, \tag{3.31}$$

and then we assign the sample $\boldsymbol{x}_n$ to the cluster s_i, even if it may be assigned to more than one cluster.

- Step 2: Update the new centroids of the new clusters formulated in iterative round r relying on:

$$\boldsymbol{\mu}_i^{(r+1)} = \frac{1}{|s_i^{(r)}|} \sum_{\boldsymbol{x}_j \in s_i^{(r)}} \boldsymbol{x}_j, \tag{3.32}$$

where $|s_i^{(r)}|$ denotes the number of samples in cluster s_i in iterative round r.

Convergence is regarded as attainable when the assignment in Step 1 is stable. Explicitly, reaching convergence means that the clusters formulated in the current round are the same as those formed in the last round. Because it is a heuristic algorithm, there is no guarantee that it can converge to the global optimum. Hence, the result of clustering largely relies on the initial clusters and their centroids.

3.2.2 EM Algorithms

The EM algorithm[2] is an iterative method to search the maximum likelihood estimate of parameters in a statistical model. Typically, in addition to unknown parameters, the statistical model also relies on some unknown latent variables, where it is difficult to achieve a closed-form solution by just taking the derivatives of the likelihood function with respect to all the unknown parameters and latent variables. As an iterative algorithm, EM algorithm consists of two steps. In expectation step (E-step), it calculates the expected value of the log likelihood function conditioned on given parameters and latent variables, while in maximization step (M-step), it updates the parameters by maximizing the expectation function log-likelihood considered.

Considering a statistical model with observable variables X and latent variables Z, the unknown parameters are represented by θ. The log-likelihood function of the unknown parameters is given by:

$$l(\theta; X, Z) = \log p(X, Z; \theta).\tag{3.33}$$

Hence, the EM algorithm can be conceived as follows:

- E-step: Calculate the expected value of the log likelihood function under the current estimate of $\bar{\theta}$, i.e.

$$Q(\theta|\bar{\theta}) = E_{Z|X,\bar{\theta}}[\log p(X, Z; \theta)].\tag{3.34}$$

- M-step: Maximize Equation (3.34) with respect to θ for achieving an updated estimate of $\bar{\theta}$, which can be given by:

$$\theta' = \arg\max_{\theta} Q(\theta|\bar{\theta}).\tag{3.35}$$

The EM algorithm plays a critical role in parameter estimation for some useful statistical models, such as the Gaussian mixture model (GMM), hidden Markov model (HMM), etc. which are beneficial for clustering and prediction.

---

[2]This sub-section is suggested to read after the introduction of estimation in Chapter 6.

3.2.3 Principal Component Analysis

State-of-the-art big data can possibly reach billions of records and millions of dimensions. It is very unlikely for all data variables to be independent, without correlation structure among them. A challenge for data scientists arises to against *multicollinearity* that some predictor variables are strongly correlated with each other, which leads to instability in solutions and inconsistent results. Bellman first noted that the sample size to fit a multivariate function grows exponentially with the number of variables [4], which suggests high dimensional data spaces are inherently sparse. Too many predictor variables not only unnecessarily complicate the analysis but also lead to overfitting. To alleviate such a technology challenge, *dimension reduction* or *feature extraction* emerges as the most prominent problem in big data analytics.

Among ways of dimension reduction for data, *principal components analysis* (PCA) serves a benchmark role. PCA interprets the correlation structure of predictor variables by a smaller set of linear combinations of these predictor variables. Such linear combinations are called *components*. In other words, the development of the concept for PCA is based on the eigenvalues/eigenvectors of the data matrix.

Suppose the original (data) variables $X_1, \ldots, X_m$ form a coordinate system in m-dimensional space. The principal components represent a new coordinate system by rotating the original coordinate system along the direction of maximum variability. Prior to dimension reduction, by appropriate data transform, the mean of each variable is set to be zero with unitary variance. That is, let each data variable X_i be $n \times 1$ vector, and

$$Z_i = \frac{X_i - m_i}{\sigma_{ii}} \tag{3.36}$$

where m_i denotes the mean of X_i and σ_{ii} denotes the standard deviation of X_i. In other words,

$$\mathbf{Z} = (\mathbf{V}^{1/2})^{-1}(\mathbf{X} - \mathbf{m}) \tag{3.37}$$

where

$$\mathbf{V}^{1/2} = \begin{bmatrix} \sigma_{11} & 0 & \cdots & 0 \\ 0 & \sigma_{22} & \cdots & 0 \\ \vdots & \vdots & \ddots & \vdots \\ 0 & 0 & \cdots & \sigma_{mm} \end{bmatrix} \tag{3.38}$$

Then, denote $\mathbf{C}$ as the symmetric covariance matrix.

$$\mathbf{C} = \begin{bmatrix} \sigma_{11}^2 & \sigma_{12}^2 & \cdots & \sigma_{1m}^2 \\ \sigma_{21}^2 & \sigma_{22}^2 & \cdots & \sigma_{2m}^2 \\ \vdots & \vdots & \ddots & \vdots \\ \sigma_{m1}^2 & \sigma_{m2}^2 & \cdots & \sigma_{mm}^2 \end{bmatrix} \tag{3.39}$$

where

$$\sigma_{ij}^2 = \frac{1}{n} \sum_{l=1}^{n} (x_{li} - m_i)(x_{lj} - m_j) \tag{3.40}$$

The purpose of covariance is to measure how two variables vary together. Please note that two independence implies being uncorrelated, but being uncorrelated does not imply independence. The correlation coefficient is

$$\rho_{ij} = \frac{\sigma_{ij}^2}{\sigma_{ii}\sigma_{jj}} \tag{3.41}$$

The correlation matrix is defined as $\mathbf{R} = [\rho_{ij}]_{m \times m}$. Considering the standardized data matrix $\mathbf{Z}$, we have

$$\mathbb{E}[\mathbf{Z}] = \mathbf{0}_{n \times m} \tag{3.42}$$

and thus

$$Cor(\mathbf{Z}) = (\mathbf{V}^{1/2})^{-1} \mathbf{C} (\mathbf{V}^{1/2})^{-1} \tag{3.43}$$

Remark: For the standardized data matrix, the covariance matrix and the correlation matrix are the same.

The ith principal component of the standardized data matrix, $\mathbf{Z} = [Z_1, \ldots, Z_m]$ which has eigenvectors $\mathbf{e}_1, \mathbf{e}_2, \ldots$, is given by

$$Y_i = \mathbf{e}_i^T \mathbf{Z} \tag{3.44}$$

where $\mathbf{e}_i$ is the ith eigenvector of $\mathbf{Z}$. The principal components, $Y_1, Y_2, \ldots, Y_k, k \leq m$, are linear combinations of the standardized data matrix $\mathbf{Z}$, such that

- the variance of the $\{Y_i\}$ are as large as possible.
- $\{Y_i\}$ are uncorrelated.

The first principal component is the linear combination

$$Y_1 = \mathbf{e}_1^T \mathbf{Z} = \sum_{l=1}^{m} \epsilon_{l1} Z_l \tag{3.45}$$

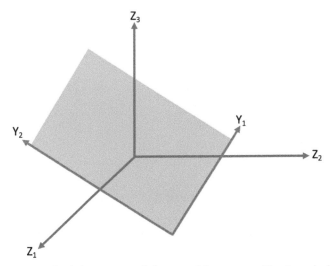

Figure 3.8   Standardized data space and the space/plane spanned by the principal components.

and has the greater variability than any other possible linear combinations. In other words, the first principal component plays the most important role with the following properties:

- The first principal component presents the linear combination $Y_1 = \mathbf{e}_1^T \mathbf{Z}$ to maximize $Var(Y_1) = \mathbf{e}_1^T \mathbf{R} \mathbf{e}_1$.
- The ith principal component presents the linear combination $Y_i = \mathbf{e}_i^T \mathbf{Z}$ that is independent of all other principal components $Y_j, j < i$ and maximizes $Var(Y_i) = \mathbf{e}_i^T \mathbf{R} \mathbf{e}_i$.

Example: Figure 3.8 illustrates a simplified example for 2-dimensional principal components derived from 3-dimensional data space.

Proposition: Suppose λ_i is the eigenvalue corresponding to the eigenvector $\mathbf{e}_i$.

(a) (Total Variability)

$$\sum_{i=1}^{m} Var(Y_i) = \sum_{i=1}^{m} Var(Z_i) = \sum_{i=1}^{m} \lambda_i = m \qquad (3.46)$$

(b) (Partial Correlation)

$$Corr(Y_i, Z_j) = e_{ij}\sqrt{\lambda_i}, i, j = 1, \ldots, m \qquad (3.47)$$

where $(\lambda_1, \mathbf{e}_1), \ldots, (\lambda_m, \mathbf{e}_m)$ are the eigenvalue-eigenvector pairs of correlation matrix $\mathbf{R}$, and $\lambda_1 \geq \lambda_2 \geq \cdots \geq \lambda_m$

(c) The proportion of the total variability in $\mathbf{Z}$ that is accounted by λ_i/m.

Remark: From the computational aspect, once the first principal component is obtained, it is more difficult to get the next few principal components.

▶ **Exercise:** Niklas participates the varsity of golf and find the heights (in cm, X) and weights (in kg, Y) of all twelve team members as follows:

Height	182	188	178	180	179	171	167	195	188	175	183	186
Weight	75	77	72	81	71	64	65	96	83	76	86	79

To infer $\mathbb{E}[Y \mid X]$ using the criterion of least squares,

(a) please find the first PCA component.
(b) please find the linear regression and compare with (a).

3.3 Deep Neural Networks

Human brain is probably the most efficient computing machine to act and to infer. The brain is a highly complex, nonlinear, and parallel computer, but quite different modern computer architecture based on digital logic circuits and bus structure, to organize *neurons* performing computations such as pattern recognition, perception, and control. *Artificial neural networks* (ANN) execute a set of algorithms on information processing hardware by imitating the interactions between neurons in human brain, which are designed to extract features for clustering and classifying. In a common ANN model, the input of each artificial neuron is real-number signals, and the output of each artificial neuron is calculated by a non-linear function of the sum of its inputs. The following three elements are used to model a neuron:

(a) A set of *synapses*, which can be represented by connecting links. Among these links, signal $x_l, l = 1, \ldots L$ as the input of the l-th synapse with weight w_{kl}.

(b) An adder, which sums the weighted input signals according to respective synaptic strengths. A bias b_k can be applied to adjust the net input of the activation function.

(c) An *activation function*, $\psi(\cdot)$, which is also known as a *squashing function* to limit the permissible amplitude range of the output signal.

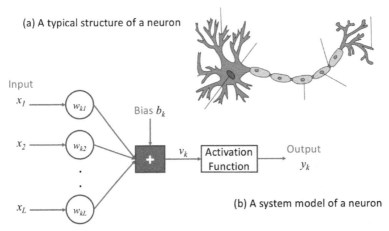

Figure 3.9 Nonlinear Model of the k-th Neuron: (a) a typical structure of a neuron from Wikipedia (b) a system model.

Mathematically,

$$y_k = \psi\left(\sum_{l=1}^{L} w_{kl}x_l + b_k\right) = \psi(u_k + b_k) = \psi(v_k) \qquad (3.48)$$

where u_k is the output of linear combiner and v_k represents the *induced local field*.

Artificial neurons and their connections typically use a weight to adjust the strength of learning process. There are ways to model the activation function:

- Threshold Function, where a typical version is the *heaviside function* as

$$\psi(v) = \begin{cases} 1, & v \geq 0 \\ 0, & v < 0 \end{cases} \qquad (3.49)$$

- Sigmoid Function, which gives one-to-one S-shape nonlinearity. A common example is the *logistic function* defined by

$$\psi(v) = \frac{1}{1 + e^{-cv}} \qquad (3.50)$$

where c is the slope parameter of the sigmoid function.

Above activation functions only define value in the range of $(0, 1)$. Sometimes, we may want to extend the range to $(-1, 1)$ and thus (3.49) can be substituted by

$$\psi(v) = \begin{cases} 1, & v \geq 0 \\ 0, & v = 0 \\ -1, & v < 0 \end{cases} \qquad (3.51)$$

To extend the scope of (3.50), we may take advantage of the hyperbolic tangent function

$$\psi(v) = \tanh(v) \qquad (3.52)$$

Moreover, artificial neurons are organized in the form of layers. Different layers perform different kinds of transformations of their inputs. Basically, signals travel from the first layer to the last layer possibly via multiple hidden layers. The DNN is a kind of deep artificial neural networks characterized with multiple hidden layers between the input and output layers. DNN is able to model complex relationships of date with the aid of multiple non-linear transformations. In a DNN, extra layers enable the composition of features from lower layers, which is beneficial in terms of modeling complex data with fewer units than a shallow network having one hidden layer. Furthermore, DNN is a type of feed-forward network, where data flows in the direction from the input layer to the output layer without looping back.

By contrast, the recurrent neural network (RNN) is a sort of artificial neural networks where any neuron in one layer is capable of connecting to the neurons in previous layers. In other words, RNN is able to exploit the dynamic temporal information for a time sequence and can substantially make use of the "memory" from previous layers to process the future inputs. Common algorithms used to training the RNN include the real time recurrent learning causal recursive backpropagation algorithm, backpropagation through time algorithm, etc.

The convolution neural network (CNN) is a class of feed-forward deep artificial neural networks with the weight-shared architecture and translation invariance characteristic, which hence requires the minimal preprocessing. In a basic CNN architecture, it is composed of an input layer, an output layer as well as multiple hidden layers, which are often consist of convolutional layers, pooling layers and fully connected layers. Particularly, the convolutional layers invoke a convolution operation, also called the cross-correlation operation to the input, and yield a multi-dimensional feature map relying on the number of filters contained. The CNN has been successfully used in image and video recognition, natural language processing, recommender systems, etc. Figure 3.10 shows the basic architecture of DNN, RNN and CNN, respectively.

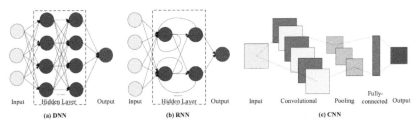

Input   Hidden Layer   Output   Input   Hidden Layer   Output   Input   Convolutional   Pooling   Fully-connected   Output

(a) DNN   (b) RNN   (c) CNN

Figure 3.10   The basic architecture of DNN, RNN, CNN [8].

Deep learning can be applied together with other machine learning techniques, such as AlphaGo©.

3.4 Data Preprocessing

Once we obtain a raw dataset from sensors or data collection, before data analysis, we must *correct* the dataset due to missing values, outliers in the data, infeasibility or inconsistency for some data in the dataset due to considerations in data fields, models, policy. Consequently, *data cleaning* is required, which actually takes a big effort in data analytics, particularly noisy sensor data in robotics.

Example (Missing Data): Suppose we are collecting the GPS dataset of vehicles for the purpose of mobility pattern prediction. One vehicle is moving through the Lincoln tunnel as Figure 3.11. Obviously, there will be some missing values in the collected GPS dataset for this vehicle due to impossible receiving GPS signals from satellites. How can we supply the missing data?

Solution: Linear interpolation is the immediate solution. Furthermore, modern vehicles usually equip with speed odometer and gyroscope, which assist further precision to fill in missing GPS data.

Another major issue in data cleaning is to identify *outliers* in a dataset, which might make our statistics much less meaningful. To illustrate, Figure 3.12 provides 25 models of vehicles (sedans and SUVs) relying on gasoline engines, with average gas mileages after testing of 10,000 miles. We may observe two unusual cases to be viewed as outliers:

- A vehicle of only 200 lbs has gas mileage at 8 mpg. Other vehicles have weights around 2,000-5,000 lbs.
- Another vehicle weighs over 5,000 lbs but has gas mileage over 50 mpg, quite different from the trend.

Figure 3.11   Map of lincoln tunnel between manhattan and New Jercy.

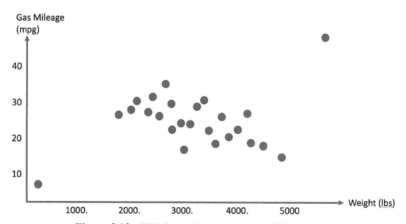

Figure 3.12   Weights and gas-mileages of 25 vehicles.

However, above observations are based on human intelligence. Is there any scientific way to identify outliers in a dataset? Furthermore, can we automatically transform outliers to normalized data (i.e. data transformation) so that dis-favorable inference will not be made? We may intuitively look into sample mean and sample variance (and thus sample standard deviation). More sophisticated methods are introduced in this section.

Proposition (Min-Max Normalization): The min-max normalization value of a data variable X is defined as

$$X^*_{mm} = \frac{X - \min(X)}{range(X)} = \frac{X - \min(X)}{\max(X) - \min(X)} \qquad (3.53)$$

Remark: Min-Max normalization transforms data into the range of $[0, 1]$, with mid-range, $\frac{\max(X)+\min(X)}{2}$, equal to $1/2$.

Proposition (Z-Score Standardization): A sample x from a data variable X is defined to have the *Z-score* through sample mean and sample variance (and thus sample standard deviation):

$$Z(x) = \frac{x - \bar{X}}{\sigma(X)} \qquad (3.54)$$

Remark: *Z-score* is analog distribution of a data variable to normal distribution. It is well know that a value of more than three standard deviation in normal distribution has rather low probability to happen (i.e. less than 10^{-3}).

Proposition (Skewness): For a data variable X, the sample median is ν_X. The skewness of X, ξ_X, is

$$\xi_X = \frac{3(m_X - \nu_X)}{\sigma_X} \qquad (3.55)$$

Remark: For *right-skewed* data, the mean is greater than the median, and thus $\xi_X > 0$, while *left-skewed* data has $\xi_X < 0$. Real-world data is usually right skewed.

Example (US Household Income): Figure 3.13 summarizes a portion of statistics according to 2014 consensus. The mean income is \$75,738 but the median income is close to \$56,000, which demonstrates typical right-skewed data.

▶ **Exercise:** Is there any outlier in the following geographical dataset (x, y, z, t)?

$$(35.107, 126.299, 35.72, 1.003)$$
$$(35.110, 126.328, 35.43, 2.082)$$
$$(35.663, 127.019, 38.92, 5.185)$$
$$(34.988, 126.630, 36.02, 6.178)$$
$$(35.149, 126.712, 36.83, 7.284)$$

▶ **Exercise:** Hannah is given the heights (in cm) and weights (in kg) of girl weight-lifting varsity in a university. Based on the following data, should the

Income of Household	Number (thousands) [47]	Percentage	Percentile	Mean Income [47]	Mean number of earners [48]	Mean size of household [48]
Total	124,587	—	—	$75,738	1.28	2.54
Under $5,000	4571	3.67%	0	$1,080	0.20	1.91
$5,000 to $9,999	4320	3.47%	3.67th	$7,936	0.34	1.78
$10,000 to $14,999	6766	5.43%	7.14th	$12,317	0.39	1.71
$15,000 to $19,999	6779	5.44%	12.57th	$17,338	0.54	1.90
$20,000 to $24,999	6865	5.51%	18.01th	$22,162	0.73	2.07
$25,000 to $29,999	6363	5.11%	23.52th	$27,101	0.82	2.19
$30,000 to $34,999	6232	5.00%	28.63th	$32,058	0.94	2.27
$35,000 to $39,999	5857	4.70%	33.63th	$37,061	1.04	2.31
$40,000 to $44,999	5430	4.36%	38.33th	$41,979	1.15	2.40
$45,000 to $49,999	5060	4.06%	42.69th	$47,207	1.24	2.52
$50,000 to $54,999	5084	4.08%	46.75th	$51,986	1.32	2.54
$55,000 to $59,999	4220	3.39%	50.83th	$57,065	1.41	2.56
$60,000 to $64,999	4477	3.59%	54.22th	$62,016	1.46	2.64
$65,000 to $69,999	3709	2.98%	57.81th	$67,081	1.51	2.67
$70,000 to $74,999	3737	3.00%	60.79th	$72,050	1.57	2.73
$75,000 to $79,999	3484	2.80%	63.79th	$77,023	1.60	2.79
$80,000 to $84,999	3142	2.52%	66.58th	$81,966	1.63	2.79

Figure 3.13 Part of Distribution of US Household Income, from Wikipedia.

Height	165	168	178	180	173	171	167	185	188	175	183	157
Weight	55	67	72	81	71	64	58	86	83	76	86	91

last person's data be kept for future statistical inference? Please explain your reason.

Remark: As we will show later that a robot operates based on a lot of sensor data that relies on wireless communication for robot's usage, such noisy data without rapid pre-processing may hurt the accuracy of robot operation very much.

Further Reading: There are a few good books in machine learning such as [3, 5, 6]. For applications of machine learning to wireless networks, the readers might refer [7] and [8], that summarize a lot of materials in this chapter.

References

[1] P. Klimek, U. Yegorov, R. Hanel, S. Thurner, "Statistical Detection of Systematic Election Irregularities", *Proceeding of National Academy of Science*, vol. 109, no. 41, pp. 16469–16473, October 9, 2012.

[2] V. N. Vapnik, "An overview of statistical learning theory", *IEEE Transactions on Neural Networks*, vol. 10, no. 5, pp. 988–999, May 1999.

[3] T. Hastie, R. Tibshirani, J. Friedman, *The Elements of Statistical Learning*, 2nd edition, Springer, 2009.

[4] R. Bellman, *Adaptive Control Processes: A Guided Tour*, Princeton University Press, 1961.

[5] C.M. Bishop, *Pattern Recognition and Machine Learning*, Springer, 2006.

[6] K.P. Murphy, *Machine Learning: A Probabilistic Perspective*, MIT Press, 2012.

[7] C. Jiang, H. Zhang, Y. Ren, Z. Han, K.C. Chen, L. Hanzo, "Machine Learning Paradigms for Next-Generation Wireless Networks", *IEEE Wireless Communications*, vol. 24, no. 2, pp. 98–105, April 2017.

[8] J. Wang, C. Jiang, H. Zhang, Y. Ren, K.-C. Chen, L. Hanzo, "Thirty Years of Machine Learning: The Road to Artificial Intelligence Pareto-Optimal Wireless Networks", *IEEE Communications Surveys and Tutorials*, vol. 22, no. 3, pp. 1472–1514, 3Q 2020.

4

Markov Decision Processes

The essential functionality of a robot or any AI agent is to make decisions or to take actions. A further challenge is that a robot usually has to take an action or a series of actions under uncertainty. As uncertainty can be usually treated by the probabilistic approach, in this chapter, the statistical framework of decision making will be introduced, first in the form of one-shot decision, and then a sequential decision process, particularly Markov decision process underlying in a dynamic system.

4.1 Statistical Decisions

Decision is one of the most fundamental human intellectual behaviors, while mathematicians pioneered by D. Bernoulli have spent over 300 years to analytically comprehend the decision processes. Modern decision theory and statistical decision theory [1] had been established in the 20th century, which has been successfully applied to communications and signal processing [2]. Let us use the following example to illustrate the uncertain in decision making process.

Example: Patrick is a graduate student driving to the university for a final examination starting at 9 am and ending at 11 am. Unfortunately, when he arrives the parking lot beside the classroom at 9 am, the parking lot is full. Patrick faces two alternatives

(a) Driving to the remote parking lot and running to classroom, which results in 20 minutes late in the final examination to likely result in poor grade

(b) Illegal parking beside the curb, which is subject to fine of $200, but the campus parking staff check once every 3 hours in average

How should Patrick decide? It appears related to the prior knowledge and cost structure.

In early days of statistics, *statistical inference* is typically done by frequentist inference, typically through the argument "what if the experiment repeats many times". Of course, Patrick can not repeat his dilemma many times in above example, and a more systematic approach is required. To extend the scope, the purpose of decision theory is trying to assist rational and logic decisions by humans and by machines. Decision theory shall therefore be able to answer the following questions:

- What is a decision?
- What is a good decision?
- How can we formally evaluate decisions?
- How can we formulate the decision problem confronting a decision maker?

Bayes theorem in the statistical inference is extremely useful to infer a decision.

Bayes Theorem: If X, Y are random variables, then

$$P(Y \mid X) = \frac{P(X \mid Y)P(Y)}{P(X)} \tag{4.1}$$

▶ **Exercise (Monte Hall):** In a popular TV show, the winner of game among attendees can participate a contest for cash award. The rules are as follows: The winner can select one of the three doors, while there is a big cash prize behind one door, and THANK YOU behind other two doors.

- Once the winner selects one door, the TV show host opens one of the other two doors and shows THANK YOU sign. The host will ask the winner whether to change his/her selection. Obviously, the host knows which door behind has the prize. Please determine the best policy for the winner, to change selection or not.
- Once the winner selects one door, an earthquake happens and thus one of the other two doors randomly opens to reveal a THANK YOU sign. The host will ask the winner whether to change his/her selection. Please determine for the winner, to change selection or not.

Proposition (Bayes Theorem of Observation Data): Let θ denote the unknown parameter; $\mathcal{D}$ denote the observed data; and $\mathcal{H}$ denote the overall hypothesis space. Then,

$$P(\theta \mid \mathcal{D}, H) = \frac{P(\mathcal{D} \mid \theta, \mathcal{H})P(\theta \mid \mathcal{H})}{P(\mathcal{D} \mid \mathcal{H})} \tag{4.2}$$

Remark: In high-level language, above proposition means

$$posterior = \frac{likelihood \times prior}{evidence}$$

In addition to AI and machine learning, decision theory also serves as the foundation of modern digital communications. The simplest example of modern digital communications can be summarized as the following illustration.

Example: A pair of transmitter and receiver uses a commonly agreed binary signal set $\{s_0(t), s_1(t)\}$ to represent "0" and "1" respectively. The transmitter sends the selected waveform into the channel to represent the binary signal to be transmitted. The channel can be wireless medium or optical fiber, with embedded noise corrupting the waveform. The receiver picks up the received waveform, $r(t)$, to determine which one of the two hypotheses to be true:

$$H_0 : r(t) = s_0(t) + n(t) \tag{4.3}$$
$$H_1 : r(t) = s_1(t) + n(t) \tag{4.4}$$

If the receiver selects H_0, it decodes the received waveform as signal "0". Similarly, if the receiver selects H_1, it decodes the received waveform as signal "1". The receiver acts as an AI agent to intelligently determine the signal based on the received waveform. Figure 4.1 illustrate this technical problem, how can we design a digital communication receiver to determine the signal bits to be transmitted based on the received waveform and the agreed signal waveforms. The receiver or the agent has to test two hypotheses

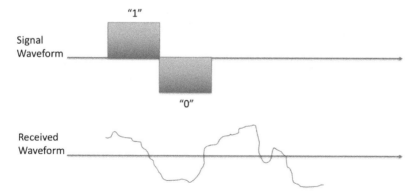

Figure 4.1 Signal detection/decision in a binary digital communication system.

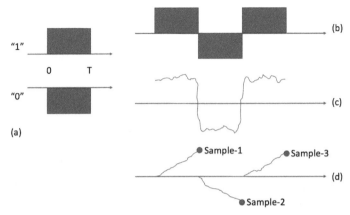

Figure 4.2   Operation principle of the matched filter (a) An anti-podal binary signal set (b) the transmitted waveform for 3 bits "101" (c) the received waveform (d) using signal waveform "1" to matched the received waveform to obtain 3 sample values.

in above equations toward the decision, and this is known as *hypothesis testing* in statistics.

Remark: A simple and intuitive realization for such hypothesis testing is *matched filter*. The rationale of matched filter is to design a receive filter to maximize the signal-to-noise ratio (SNR). Under the additive white Gaussian noise (AWGN) which is widely applicable even in many pattern recognition problems, the impulse response of matched filter is

$$h(t) = S^*(T - t), 0 \leq t \leq T \qquad (4.5)$$

where $s_1(t) = -s_0(t) = S(t), 0 \leq t \leq T$ and $*$ denotes complex conjugate. Figure 4.2 illustrates the operation of matched filter. The impulse response of matched filter in this case is simply the time-reverse waveform of $s_1(t)$ that remains the same. Assuming perfect synchronization, the received waveform is fed into the matched filter and is sampled once every T then reset, which gives 3 sample values. It is intuitive (actually mathematically true for equally probable cases) that the decision threshold can be set to zero due to antipodal signal waveforms. For sample values larger than 0, "1" is determined. Otherwise, "0" is determined, to complete a statistical decision mechanism. The same principle can be generalized to other *statistical pattern recognition* problems such as Figure 3.6.

Remark: Hypothesis testing is equivalent to a wide range of AI decision problems in addition to digital communications, such as classification, pattern recognition, etc. More generally, *statistical inference* involves different classes of problems: *hypothesis testing, estimation, prediction,* and *ranking*.

4.1.1 Mathematical Foundation

It is a natural behavior for human beings to make a decision based on earlier experience that could often be modeled as statistics due to the uncertainty. Statistical Decision Theory is the mathematical framework to make decisions in the presence of statistical knowledge. Classical statistics uses sample information to directly infer parameter θ. (Modern) Decision Theory makes the best decision by bringing sample information with relevant aspects of the problem, with the following two ingredients brought up in the 1960's

- possible consequence of the decision introduced by Abraham Wald
- *a prior* information introduced by L.J. Savage

Consequently, a decision problem involves a set of states, S, a set of potential consequences of decisions, $\mathcal{X}$. A decision (or known as an *act* or *action*) is considered as a mapping from the state space to the consequence set. That is, in each state $s \in S$, an act a generates a well-defined consequence $a(s) \in \mathcal{X}$. The decision maker must rank acts without precise knowing current state of the world. In other words, an act is conducted with uncertainty.

The consequences of an act can often be ranked in terms of relative appeal, that is, some consequences are better than others. Thanks to Von Neumann and Morgenstein in pioneer developing game theory, this is often numerically modeled via a *utility* function, u, which assigns a utility value $u(x) \in \mathbb{R}$ for each consequence $x \in \mathcal{X}$. To model the uncertainty of knowledge about the states, we usually assume a probability distribution π over S as the *a prior* information, which can be obtained by empirical or statistical methods in advance.

Usually, a decision rule proceeds on selecting an action to result in the preferred expected utility

$$\mathbb{E}[U] = \sum_{s \in S} \pi(s)u[a(s)] \tag{4.6}$$

Of course, we usually prefer larger utility, which leads to maximization of utility. In many cases, we consider the cost, risk, or loss of an action, which suggests minimization of the cost, for example, to design a highly reliable system like a robot or a communication receiver.

Example: A digital communication system requires extremely high accuracy in signal detection. Bit error rate (BER) or probability of error is usually used as a measure of receiver. An error is treated as the cost of an action. A common requirement of BER is 10^{-6}, which means one error in average among one million decisions at the receiver.

A *decision space* of M elements (or distinct decisions) can be defined as $\mathcal{A} = \{a_0, a_1, \ldots, a_{M-1}\}$. A decision rule $\delta(s) = a \in \mathcal{A}$ maps the state to an action by evaluating utility. A non-randomized devision rule selects one action from $\mathcal{A}$. A randomized decision rule determines the probability distribution over the elements of $\mathcal{A}$, that is, $p_\delta[\delta(x) = a_m], m = 0, \ldots, M-1$.

Finally, appropriate decision principle must be selected. If both the cost function and *a priori* probability, $\pi(\cdot)$ are known, *Bayes* decision rule is appropriate. If the cost function is known but not *a priori* probability, *minimax* decision rule shall be used. Finally, if neither cost function nor *a priori* probability is known, *Neyman-Pearson* decision rule could be employed.

4.1.2 Bayes Decision

Bayesian decisions are made based on the known priori probability and cost (or equivalent terms such as risk and loss). We use the following example to illustrate how Bayesian decision proceeds.

Example (Binary Hypothesis Testing): Suppose we intend to detect an electric signal on a cable by measuring the voltage. For each measurement, the empirical experience suggests *a priori* probability to have signal to be $1/2$, that is, $\pi_1 = \pi_0 = 1/2$. If the signal is present, the measured voltage is expected to be A, a positive constant. Otherwise, the measured voltage is expected to be 0. However, an additive Gaussian noise with zero mean and variance σ^2 is embedded to disturb the measurement, that is, $n \sim G(0, \sigma^2)$. The cost to make an erroneous detection is a constant C_b. This exactly forms a binary *hypothesis testing* problem as follows:

$$H_1 \ (signal \ presence) : \ y = A + n \tag{4.7}$$

$$H_0 \ (signal \ absence) : \ y = n \tag{4.8}$$

The subsequent *likelihood functions* are

$$f_1(y \mid H_1) = \frac{1}{\sqrt{2\pi\sigma^2}} e^{-\frac{1}{2}(y-A)^2} \tag{4.9}$$

$$f_0(y \mid H_0) = \frac{1}{\sqrt{2\pi\sigma^2}} e^{-\frac{1}{2}y^2} \tag{4.10}$$

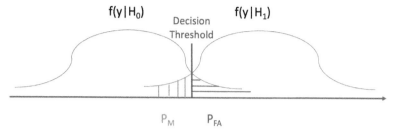

Figure 4.3 Likelihood functions for two hypotheses.

As shown in Figure 4.3, the decision mechanism is simply to determine which hypothesis is more likely. More precisely, it is equivalent to comparing $\pi_1 f_1(y \mid H_1)$ and $\pi_0 f_0(y \mid H_0)$ in order to determine which is larger.

We usually formulate as the following *likelihood ratio test*:

$$L(y) = \frac{\pi_1 f_1(y \mid H_1)C_b}{\pi_0 f_0(y \mid H_0)C_b} \overset{H_1}{\underset{H_0}{\gtrless}} 1 \tag{4.11}$$

To remove the exponential operation, we introduce logarithm that is a one-to-one function to preserve the relationship in comparison, and thus form the *log likelihood ratio test* as

$$l(y) = \log L(y) = Ay - \frac{1}{2}A^2 \overset{H_1}{\underset{H_0}{\gtrless}} 0 \tag{4.12}$$

where we use $\pi_1 = \pi_0 = 1/2$. It suggests a simple decision mechanism after obtaining the noisy measurement y, to determine

$$H_1 : if \ y > \frac{A}{2}$$

$$H_0 : if \ y < \frac{A}{2}$$

where we can use randomized or arbitrary decision if $y = A/2$.

The probability of error in this binary decision/detection is

$$P_e = \pi_1 P(H_0 \mid H_1) + \pi_0 P(H_1 \mid H_0)$$

$$= \frac{1}{2}Q\left(\frac{A/2}{\sigma}\right) + \frac{1}{2}Q\left(\frac{A/2}{\sigma}\right) = Q\left(\frac{A/2}{\sigma}\right) \tag{4.13}$$

where $Q(x) = \int_x^\infty \frac{1}{\sqrt{2\pi}}e^{-t^2/2}dt$ is the Gaussian tail function.

Remark: This methodology can be easily extended into M candidates in decision, with known *a priori* distribution.

Above example is one-dimensional data. To deal with high-dimensional data (e.g. the pattern recognition exercise in Chapter 3), we consider the following M-ary hypothesis testing:

$$H_i : y(t) = s_i(t) + n(t), \ i = 1, \ldots, M \tag{4.14}$$

Suppose we have an orthonormal functional basis $\{\phi_k(t)\}_{j=1}^N$, which means

$$\int_0^\infty \phi_i(t)\phi_j(t)dt = \delta_{ij} \tag{4.15}$$

Consequently,

$$n_i = \int_0^\infty n(t)\phi_i(t)dt \tag{4.16}$$

Lemma: Suppose $n(t)$ is a white Gaussian noise with zero mean and power spectral density $N_0/2$. Then,

(a) $\{n_i\}$ are jointly Gaussian
(b) $\{n_i\}$ are mutually uncorrelated and thus mutually independent
(c) $\{n_i\}$ are independent and identically distributed Gaussian with zero mean and power spectral density $N_0/2$

Similarly, we can expand $s_i(t)$

$$s_{ij} = \int_0^\infty s_i(t)\phi_j(t)dt \tag{4.17}$$

Therefore, $\mathbf{s_i} = (s_{i1}, s_{i2}, \ldots, s_{iN})^T$. We successfully transform a continuous function to a signal vector in N-dimensional *signal space*. For pattern classification or pattern recognition problems, signal space is a useful concept, while signal detection here is a kind of classification problem.

Remark: If we just know $s_i(t)$, $i = 1, \ldots, M$ but not $\{\phi_k(t)\}_{j=1}^N$, we can use *Gram-Schmidt* procedure to obtain a orthonormal functional basis.

For observation signal $y(t)$, we also have

$$y_i = \int_0^\infty y(t)\phi_i(t)dt \tag{4.18}$$

to form $\mathbf{y}$. Then, the likelihood functions

$$f_Y(\mathbf{y} \mid H_i) = \prod_{j=1}^N f_{Y_j}(y_j \mid H_i), \quad i = 1, \ldots, M \tag{4.19}$$

where the product comes from independence. Due to AWGN,

$$f_Y(\mathbf{y} \mid H_i) = (2\pi \frac{N_0}{2})^{-N/2} \exp\left[-\frac{1}{2\frac{N_0}{2}} \sum_{j=1}^{N} (y_i - s_{ij})^2 \right] \qquad (4.20)$$

The consequent log likelihood function

$$l(H_i) = \left[-\frac{1}{N_0} \sum_{j=1}^{N} (y_i - s_{ij})^2 \right] \qquad (4.21)$$

The decision is made by

$$\hat{i} = \underset{i}{\operatorname{argmax}}\, l(H_i) \qquad (4.22)$$

Since

$$\sum_{j=1}^{N} (y_i - s_{ij})^2 = \|\mathbf{y} - \mathbf{s_i}\|^2 \qquad (4.23)$$

then the decision is made by selecting

$$\hat{i} = \underset{i}{\operatorname{argmin}} \|\mathbf{y} - \mathbf{s_i}\|^2 \qquad (4.24)$$

which means that the decision is made by identifying $\mathbf{s_i}$ closest (in squared Euclidean distance) to observation $\mathbf{y}$. Figure 4.4 illustrates this concept in the signal space. Since the observation $\mathbf{y}$ is closest to $\mathbf{s_i}$ in Euclidean distance (i.e. red line in the figure), H_i is therefore selected, which implies signal S_i is transmitted.

Remark: Above framework generally holds for statistical pattern recognition or signal classification/detection.

▶ **Exercise:** The meteorites can come from outer space or inside the solar system, with equally probable chances. The only difference is the radiation. According to statistics, the strength of radiation is randomly distributed among meteorites from solar system in Poisson distribution with parameter λ_0. Similarly, the strength of radiation is randomly distributed among meteorites from outer space in Poisson distribution with parameter $\lambda_1 > \lambda_0$. Please design a robot to distinguish a newly collected meteorite from out space or inside solar system, based on the measurement of radiation.

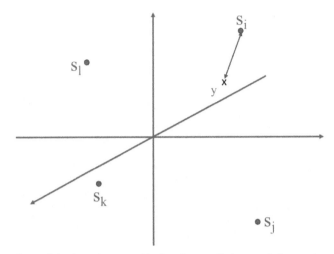

Figure 4.4   Signal space with signal constellations and observation.

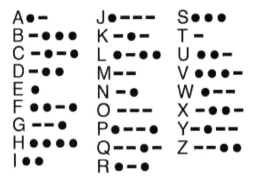

Figure 4.5   Morse codes in long (in −) and short (in •) transmissions.

▶ **Exercise:** Suppose we use Morse codes to transmit an English letter with equal probability. Eugene listens the radio and take note the received sound as • • − −, while there exists a small probability to make a mistake including (i) mistaking • by −, or vise versa (ii) the fourth transmission symbol does not exist but being record. Both cases are independent. The probability to make an error as case (i) is $0 < \epsilon < 1/2$ and the probability to make an error as case (ii) is 2ϵ. Please determine which letter is most likely to be transmitted based on Eugene's record.

Remark: In case the priori probability $\pi(\cdot)$ is not fully known, say a dummy variable θ, which is not unusual in engineering, the following approach is common to develop *generalized likelihood ratio test* (GLRT) to make Bayesian decisions. We first select the *least favorable distribution* $w_i(\theta), i = 0, 1$ of the parameter θ in decision, then form the GLRT for binary scenario in terms of additive noise distribution

$$L(y) = \frac{\int f_n(y - s_1(\theta))w_n(\theta)d\theta}{\int f_0(y - s_0(\theta))w_0(\theta)d\theta} \tag{4.25}$$

One common realization of the *least favorable distribution* is based on the principle of *maximum entropy*, while uniform distribution is one example to maximize entropy according to *information theory*.

▶ **Exercise (Composite Decision):** The binary frequency shifted keying (BFSK) modulation uses the binary signals as follows to form a binary detection. In case θ is known, a coherent demodulation/detection is realizable and thus a detection of equally probable orthogonal signaling $0 \leq t \leq T$

$$\Omega_0 = \{A\sqrt{2}\cos(2\pi f_0 t + \theta) \tag{4.26}$$
$$\Omega_1 = \{A\sqrt{2}\cos(2\pi f_1 t + \theta) \tag{4.27}$$

where $0 \leq t \leq T$ and $|\ f_1 - f_0\ | \gg 1$. Non-coherent detection of BFSK as binary *composite hypothesis testing* proceeds where θ is an unknown random variable that is usually assumed to be uniformly distributed over $[0, 2\pi]$. The sample space in this scenario is not binary, which actually consists of infinite number of points to form two subsets. By averaging over θ, we can construct a Generalized Likelihood Ratio Test to build up the optimal receiver. Please follow the following steps (1) construct the signal space (2) identify least favorable distribution of θ (3) finalize the decision rule (4) calculate energy per bit E_b (5) show the average probability of error under additive white Gaussian noise (AWGN) with zero-mean and 2-side power spectral density $N_0/2$, given error as the cost, to be $P_{e,BFSK} = Q(\sqrt{\frac{E_b}{N_0}})$.

Remark: In case we know the cost function but not *a priori* probability distribution of hypotheses, it is generally a *minimax* decision problem, while the solution, if existing, can be derived by *equalizer equations* or by the *least favorable distribution*. More details can be found in [2]. Or, we may estimate *a priori* probability distribution of hypotheses if enough data available. When the decisions without *a priori* probability distribution of hypotheses involve multiple agents, *game theory* is usually useful.

4.1.3 Radar Signal Detection

Invented during World War II to provide early warning of airplanes, *radio detection and ranging* (RADAR) is now a technique widely used in robotics. The principle of radar is to send a radio waveform onto certain direction. If there exists an object (say, an airplane made of metal that is good in reflection) to reflect such waveform back, the time difference between the transmission time and the receiving time, Δt, can be used to determine the distance d of the objection in that direction since $d = c(2\Delta t)$, where c is the speed of light.

Example: If we want to design a mobile robot to move along the hallway in a building, a critical functionality is to avoid hitting the walls. One possible design is to use a radar to detect the wall in front, and even to construct the tomography image by scanning the environment. In recent year, lidar emerges as an even more powerful technology like radar to act as the vision for autonomous vehicles.

The radar detection forms a binary hypothesis testing problem as follows:

$$H_1 \ (signal \ presence) : \ y = s + n \qquad (4.28)$$
$$H_0 \ (signal \ absence) : \ y = n \qquad (4.29)$$

Without loss of generality, to simplify the problem, let s be a constant representing the energy of the received radio waveform, and n be the AWGN with zero mean and variance σ^2. There are two kinds of errors:

- *false alarm* with probability of $P_{FA} = P_{01}$, also known as *type I error*, which means no signal but making a decision of signal presence.
- *missing* with probability of $P_M = P_{10}$, also known as *type II error*, which means signal presence but a decision of no signal being made.

The probability of detection is therefore $P_D = P_{11} = 1 - P_{10}$. Since it is hard to define the cost function nor to know *a priori* probability, this radar detection problem shall adopt the Neyman-Pearson criterion in decision making. The optimization problem for radar detection is therefore: for a given false alarm probability $P_{FA} = \alpha$, maximize $P_D = P_{11}$.

In other words, in this binary hypothesis testing with $\mathcal{A} = \{a_0, a_1\}$, given

$$P_{FA} = \int_{\mathcal{X}} \delta(a_1 \mid y) f(y \mid H_0) dy = \alpha, \ 0 < \alpha < 1 \qquad (4.30)$$

we intend to find a decision rule $\delta(a_1 \mid y)$ to maximize

$$P_D = \int_{\mathcal{X}} \delta(a_1 \mid y) f(y \mid H_1) dy \qquad (4.31)$$

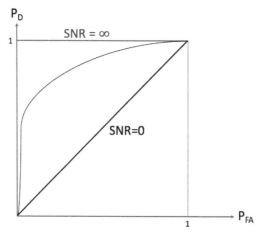

Figure 4.6   Receiver operating curve.

where $\mathcal{X}$ denotes the observation space and $y \in \mathcal{X}$.

The implementation of radar detection problem becomes a likelihood ratio test: if $\frac{f(y|H_1)}{f(y|H_0)} > K$, signal presence is claimed. It leads to, if $y \cdot s > \frac{\|s^2\|}{2} + \sigma^2 \log K = \eta$, signal presence is determined, where η is the decision threshold and can be determined from the false alarm probability.

The performance of a receiver or a detector can be represented by the *receiver operating curve* (ROC) shown in Figure 4.6. Two extreme cases (in terms of signal-to-noise ratio, SNR) are shown but the black curve indicates a common situation.

▶ **Exercise:** Radar is widely applied in robotics such that a mobile robot can sense the environment. For radar detection problem as (4.28) and (4.29), suppose $\|s\|^2 = 100$ and $n \sim G(0, 1)$. With $P_{FA} \leq 10^{-2}$, please design the radar detection mechanism.

Remark: Radar detection has to acquire reflected waveform, which usually involves some unknown parameter(s), and thus composite hypothesis and GLRT can be applied in radar detection together.

4.1.4 Bayesian Sequential Decision

Up to this moment, we discuss detection (i.e. decision or hypothesis testing) given a fixed number of samples (i.e. observations). However, a decision can be made after a few non-determined samples for the purpose of better decision-making. Such a detection or decision using a random number of samples depending on the observation sequence is known as *sequential detection* or *sequential decision* process.

Suppose independent and identically distributed (i.i.d.) observations $Y_k = \{y_k, k = 1, 2, \ldots\}$ over the observation space $\mathcal{X} = \mathbb{R}^n$ to proceed hypothesis testing over $\mathcal{P}_0, \mathcal{P}_1$ defined on $(\mathbb{R}, \mathcal{B})$. That is,

$$H_0 : Y_k \sim \mathcal{P}_0 \tag{4.32}$$
$$H_1 : Y_k \sim \mathcal{P}_1 \tag{4.33}$$

To avoid the infinite long operation, the *stopping rule* is defined as $\mathcal{O} = \{o_j, j = 0, 1, \ldots\}$, where $o_j \in \{0, 1\}^j$, and the terminal decision rule is defined as $\mathcal{D} = \{\delta_j, j = 0, 1, \ldots\}$. A *sequential decision rule* is a pair of sequences $(\mathcal{O}, \mathcal{D})$. For an observation sequence, $y_1, y_2, \ldots, y_k, \ldots$, the rule $(\mathcal{O}, \mathcal{D})$ makes the decision $\delta_N(y_1, \ldots, y_N)$, where N is known as the *stopping time* defined as $N = \min\{n \mid o_n(y_1, cdots, y_n) = 1\}$ to stop observation (or sampling) and proceed making a decision. Please note that the stopping time N is obviously random.

To derive the optimal Bayesian decision, *priors* π_0, π_1 are assigned to hypotheses H_0, H_1 respectively. Sampling each observation has a constant cost $C > 0$, and the cost to take n samples is nC. The conditional costs for a given sequential decision rule are

$$c_0(\mathcal{O}, \mathcal{D}) = \mathbb{E}_0\{\delta_N(y_1, \ldots, y_N)\} + C\mathbb{E}_0(N) \tag{4.34}$$
$$c_1(\mathcal{O}, \mathcal{D}) = 1 - \mathbb{E}_1\{\delta_N(y_1, \ldots, y_N)\} + C\mathbb{E}_1(N) \tag{4.35}$$

where the subscripts correspond to the hypotheses. The average Bayesian cost is therefore

$$R(\mathcal{O}, \mathcal{D}) = \pi_0 c_0(\mathcal{O}, \mathcal{D}) + \pi_1 c_1(\mathcal{O}, \mathcal{D}) \tag{4.36}$$

and the desired Bayesian sequential rule is to minimize the $R(\mathcal{O}, \mathcal{D})$.

As shown in Figure 4.7, the operation of sequential decision/detection proceeds as follows. Suppose $\pi_L < \pi_1 < \pi_U$, the optimum test takes at least one sample, which shall give more information about whether the hypothesis is true. Instead of *a priori* probability π_1, $\pi_1(y_1)$ serves the actual *posteriori* probability of H_1 given the observation y_1. With $y_1, \ldots, y_{n-1}$ without stopping, the optimum sequential test, after taking another observation y_n, stops and selects

- H_0 if $\pi_1(y_1, \ldots, y_n) \leq \pi_L$
- H_1 if $\pi_1(y_1, \ldots, y_n) \geq \pi_U$

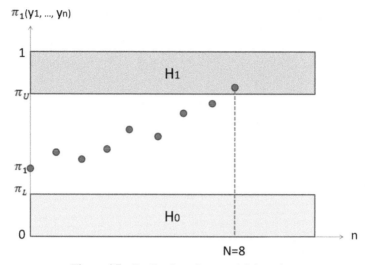

Figure 4.7 Realization of sequential detection.

- nothing but takes another sample if $\pi_L < \pi_1(y_1, \ldots, y_n) < \pi_U$

Consequently, the sequential decision is described by the stopping rule

$$\mathcal{O}_n(y_1, \ldots, y_n) = \begin{cases} 0, \; if \; \pi_L < \pi_1(y_1, \ldots, y_n) < \pi_U \\ 1, \; otherwise \end{cases} \tag{4.37}$$

and the terminal decision rule

$$\delta_n(y_1, \ldots, y_n) = \begin{cases} 1, \; if \; \pi_1(y_1, \ldots, y_n) \geq \pi_U \\ 0, \; if \; \pi_1(y_1, \ldots, y_n) \leq \pi_L \end{cases} \tag{4.38}$$

Remark: In most cases of interest, $\pi_1(y_1, \ldots, y_n)$ converges almost surely to 1 under H_1 and to 0 under H_0.

Remark: Sequential decision under proper design should have better performance than one-shot decision. Consequently, the sequential test terminates with probability 1. However, the derivations of π_L, π_U might not be easy.

4.2 Markov Decision Processes

All the decision methods in earlier section are usually considered as *one-shot decision*, except the sequential decision that makes a decision based on a period of observations. With the optimality of sequential decision, we would

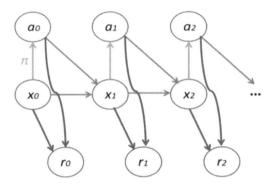

Figure 4.8 A sequential decision process with rewards.

like to explore more on the general *multi-stage decision*, which separates the decision process into a number of stages or steps and each stage usually involves the optimization of one variable. A recursive algorithm is commonly employed to compute at different stages to reach feasible optimal solution of the entire problem at the last stage. The optimization of each stage results in a decision or an action, and a sequence of decisions is therefore called a *policy*. A few states representing system status or nature are associated with each stage, to possibly influence the decision.

Furthermore, decisions may not be static in many situations, especially with uncertainty and dynamics. The outcome of decision may affect the system or the mechanism. For example, in a business transaction that user Alice is negotiating price with user Bob, if Alice can first determine whether Bob is trustworthy, price negotiation in the following can be affected. Consequently, the system state (i.e. status) can be introduced into the decision mechanism. In case the system can be modeled as a Markov chain, we have a *Markov decision process* (MDP), which serves the foundation of *reinforcement learning* in artificial intelligence.

In particular, the agent (i.e. a robot) makes a sequential decision, in which the agent immediately gets a reward (or utility, cost, etc.) from its decision to form the following scenario shown in Figure 4.8. In this chapter, we explore further into such a scenario, which is particularly useful for the purpose of system control.

4.2.1 Mathematical Formulation of Markov Decision Process

Real world decision-making often encounters multiple-objective or non-commensurate situations, decision under uncertainty and risk, or having impacts from the decision. A mathematical optimization of discrete-stage sequential decision in a stochastic environment can be developed as *Markov decision process* (MDP), via the dynamics of a controlled Markov process. The Markov process becomes a controlled Markov process when the transition probabilities can be affected by the action (i.e. decision), which has further enriched meaning than dynamic programming.

In MDPs, the interaction between the decision maker (also referred to as *agent*)and environment is described by *states*, *actions*, and *rewards*. Anything that cannot be arbitrarily changed by the agent is considered to be part of the environment. Agent observes the information pattern of the environment and establish an awareness of its own state. According to its current state, it possesses some choices of actions that will transits itself to another state and then receive some real-valued rewards from the environment. A sequence of state-action-reward happens in a so-called *decision epoch*. An MDP consists of a number of decision epochs, which we call the MDP's *horizon length*. The goal of the agent, and therefore the objective of MDP analysis, is to determine a rule for each decision epoch for selecting an action such that the collective rewards are optimized.

To form MDP in a mathematical way, let $\{X_k \in \mathcal{S}\}$ be the sequence of states at epoch $k \in \{0, 1, 2, \ldots, K\}$ where $K \leq \infty$ is the horizon length and $\mathcal{S}$ is the state space. Action taken at decision epoch k is denoted by a_k which belongs to the action space $\mathcal{A}$. More precisely, the action space can depend on the current state, denoted by $\mathcal{A}(x_k)$. The real-valued reward R_{k+1} is accrued after epoch k. Since the reward is given on the basis of last state x_k and action a_k, sometimes we use $r(x_k, a_k)$ to emphasize their relationship. The notation for reward R_{k+1} (instead of R_k) is used because it captures the fact that after certain epoch k, reward R_{k+1} and the successor state x_{k+1} are determined together.

Figure 4.9 illustrates the interaction between the agent and the environment. Obviously, we could simply describe an MDP as a sequence of

$$X_0, A_0, R_1, X_1, A_1, R_2, X_2, A_2, \ldots, R_K, S_K.$$

In the process of deciding actions, the agent follows specific decision rules $\{\delta_k\}$ corresponding to each epoch $k \in \{0, 1, \ldots, K\}$ that tells which action it should take with respect to its state. That is, the decision rule δ_k is a function mapping every $x \in \mathcal{S}$ onto an action $a \in \mathcal{A}$. In general, we refer to the sequence of decision rules as *policy* and denote it by $\pi = \{\delta_0, \ldots, \delta_K\}$. If the decision rules are the same in spite of epochs, the policy is *stationary* and we denote it by $\bar{\pi} = \{\delta, \delta, \ldots\}$.

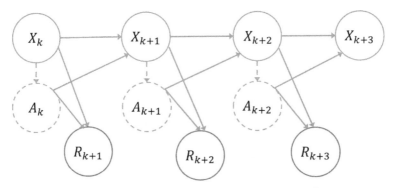

Figure 4.9   A Markov decision process from time k.

The goal of MDP analysis is to derive an optimal policy π^*, where optimal means that no matter at which state the agent is, executing the policy will lead to the maximal expected future rewards.

$$\pi^* \triangleq \arg\max_\pi v_\pi(x) \quad \forall x \in \mathcal{S} \tag{4.39}$$

where $v_\pi(x)$ is defined to be the *state-value function* (or *value function* for short) for the agent who follows policy π at state x. The criterion to measure the value is simply the expected reward in future, i.e.

$$v_\pi(x) \triangleq \mathbb{E}_\pi \left[\sum_{i=k+1}^{K} R_i \middle| X_k = x \right]$$

In case for finite horizon, $K < \infty$, the value function is certainly bounded as long as the real-valued reward $R_i < \infty$. Sometimes a *terminal rewards* $\gamma(x_{trm})$ is given when the state at the end of the MDP is some specific states, $X_K = x_{trm}$. Terminal rewards are only accrued in finite horizon MDP.

In many cases that the duration or span of the MDP execution is not clear, it is more appropriate to consider infinite horizon length. To ensure the value function converge, a discount factor $0 \le \beta < 1$ is required. Hence, a general form of the value function can be written as

$$v_\pi(x) = \mathbb{E}_\pi \left[\gamma(x_{trm}) \mathbb{1}\{K < \infty\} + \sum_{i=k+1}^{K} \beta^{i-k-1} R_i \middle| X_k = x \right] \tag{4.40}$$

where the discount factor β subjects to

$$\begin{cases} 0 \leq \beta < 1 & \text{if } K = \infty, \\ 0 \leq \beta \leq 1 & \text{if } K < \infty. \end{cases} \tag{4.41}$$

Despite state-value function, we can also evaluate state-action pairs (x, a) for policy π. Denoted by $q_\pi(x, a)$, it is referred to as *action-value function for policy π*,

$$q_\pi(x, a) \triangleq \mathbb{E}_\pi \left[\gamma(x_{trm}) \mathbb{1}\{K < \infty\} + \sum_{i=k+1}^{K} \beta^{i-k-1} R_i \,\middle|\, X_k = x, A_k = a \right] \tag{4.42}$$

with condition identical to (4.41).

4.2.2 Optimal Policies

Before looking into the value function $v_\pi(x)$, we introduce another notation G_k to denote the sum of (discounted) rewards received after epoch k.

$$G_k = \sum_{i=k+1}^{K} \beta^{i-k-1} R_i \tag{4.43}$$

Hence, the value function becomes

$$v_\pi(x) = \mathbb{E}_\pi \left[\sum_{i=k+1}^{K} \beta^{i-k-1} R_i \,\middle|\, X_k = x \right] \tag{4.44}$$

$$= \mathbb{E}_\pi \left[R_{k+1} + \beta \sum_{i=k+2}^{K} \beta^{i-k-2} R_i \,\middle|\, X_k = x \right] \tag{4.45}$$

$$= \mathbb{E}_\pi \left[R_{k+1} + \beta G_{k+1} \,\middle|\, X_k = x \right] \tag{4.46}$$

The second term in the last row is in fact related to the action taken at that epoch and the next state. If we let $\pi(a|x)$ be the probability of choosing action $A_k = a$ at state $X_k = x$; let the following reward $R_{k+1} = r$ and successor state $X_{k+1} = x'$ determined by function $p(r, x'|x, a)$, above equation turns into

$$\sum_{a \in \mathcal{A}(x)} \pi(a|x) \mathbb{E}_{\pi} \left[R_{k+1} + \beta G_{k+1} \Big| X_k = x, A_k = a \right]$$

$$= \sum_{a \in \mathcal{A}(x)} \pi(a|x) \sum_{x'} \sum_{r} p(r, x'|x, a) \left[r + \beta \mathbb{E}_{\pi}[G_{t+1}|X_{t+1} = x'] \right]$$

Consequently, we obtain

$$v_{\pi}(x) = \sum_{a \in \mathcal{A}(x)} \pi(a|x) \sum_{x'} \sum_{r} p(r, x'|x, a) \left[r + \beta v_{\pi}(x') \right] \qquad (4.47)$$

Equation (4.47) is known as the *Bellman equation* for $v_{\pi}(x)$, which implies the relationship between a state's value function and its successor state's value. Something nice to have this form of value function is that, at every decision epoch k for any state x_k, the optimal decision rule δ_k is to choose the action that will maximize the expected reward plus the discounted value of the next state.

$$\delta_k^* = \arg \max_{a \in \mathcal{A}(x_k)} \sum_{a \in \mathcal{A}(x)} \pi(a|x) \sum_{x'} \sum_{r} p(r, x'|x, a) \left[r + \beta v_{\pi}(x') \right]$$

$$= \arg \max_{a \in \mathcal{A}(x_k)} \sum_{x'} \sum_{r} p(r, x'|x, a) \left[r + \beta v_{\pi}(x') \right], \quad \forall x_k \in \mathcal{S} \quad (4.48)$$

Gathering all decision rules for each epoch constitutes an optimal policy $\pi^* = \{\delta_1^*, \ldots, \delta_K^*\}$.

4.2.3 Developing Solutions to Bellman Equation

Although we have the idea about how the optimal policy should satisfy, questions regarding the path to the exact solution still remains. But it is obvious that the only uncertainty left in equation (4.67) is the value function $v_{\pi}(\cdot)$. Once we have the precise, or say optimal, value function, the concrete formulation of π^* will emerge. More precisely, an optimal state-value function is the one in which every state's value is maximized over all policies.

$$v^*(x) = \max_{\pi} v_{\pi}(x), \quad \forall x \in \mathcal{S}.$$

Thereby, a policy is optimal if it can find out the optimal value function no matter which state the MDP starts from.

$$v_{\pi^*}(x_0) = v^*(x_0), \quad \forall x_0 \in \mathcal{S}.$$

Now, three propositions that would hopefully lead us to the solution of the MDP are introduced. The first states that there exists only one optimal value function, that is, there is a unique fixed point within the set of real-valued functions. The second and the third propositions provide basis for two general methods to obtain the optimal policy: *value iteration* and *policy iteratioin.*

Proposition 1 (Unique Solution of Bellman Equation). *Given the reward and state transition probability $p(r, x'|x, a)$, the optimal value function v^* is the unique solution of Bellman equation.*

Proposition 2 (Convergence of Value Iteration). *Any sequence $\{v_k\}$ starting from any $v_0 \in V$ and updated by*

$$v_{k+1}(x) = \max_{a \in \mathcal{A}(x)} \sum_{x'} \sum_{r} p(r, x'|x, a) \left[r + \beta v_k(x') \right], \quad \forall x \in \mathcal{S} \quad (4.49)$$

will converge to v^*.

Proposition 3 (Convergence of Policy Iteration). *Starting from any stationary policy $\bar{\pi}_0$, if the value functions are updated through*

$$v_{\bar{\pi}_i}(x) = \sum_{a \in \mathcal{A}(x)} \bar{\pi}_i(a|x) \sum_{x'} \sum_{r} p(r, x'|x, a) \left[r + \beta v_{\bar{\pi}_i}(x') \right], \quad \forall i = 1, 2, \ldots,$$

$$(4.50)$$

after the value of all states converge, improve upon the policy to construct another stationary one $\bar{\pi}_{i+1} = \{\delta_{i+1}, \delta_{i+1}, \delta_{i+1} \ldots \}$ by

$$\delta_{i+1}(x) = \arg \max_{a \in \mathcal{A}(x)} \sum_{x'} \sum_{r} p(r, x'|x, a) \left[r + \beta v_{\bar{\pi}_i}(x') \right]. \quad (4.51)$$

The sequence of value function $\{v_{\bar{\pi}_i}\}$ will converge to the unique optimal one v^*.

Remark: There are several explicit versions of proof for these propositions. However, we will just give a brief explanation on the idea in the following.

Define two operators H_δ and H for any real-valued function v on $\mathcal{S}$:

$$(H_\delta v)(x) = \mathbb{E}_\pi[R_{k+1} + \beta v(x')|X_k = x]$$

$$Hv = \sup_{\delta} \{H_\delta v\}$$

Then, the recursion equation can be rewritten as

$$v_{k+1} = Hv_k, \quad k = 0, \ldots, K - 1$$

$$\delta = \delta^*_{K-k} \text{ if and only if } H_\delta v_k = Hv_k$$

Now, we turn our attention to infinite horizon discounted problem. Let V be the set of all real-valued functions on S. $H : V \to V$. Define

$$\|v\| = \max\{|v(i)| : i \in S\}.$$

Furthermore, $\forall u, v \in V$,

$$\|Hu - Hv\| \le \beta \|u - v\|.$$

$\beta < 1$ guarantees that H is a contraction operator and thus implies H has a unique fixed point in V. That is, there exists a unique $v^* \in V$ such that $v^* = Hv^*$. (Proof of Proposition 1.) Moreover, For any sequence $\{v_k\}$ such that $v_0 \in V$ and $v_{k+1} = Hv_k$,

$$\lim_{k \to \infty} \|v^* - v_k\| = 0$$

which is the exact concept of value iteration in proposition 2.

Example (One-Armed Bandit): Matthew goes to Las Vegas for a conference. Before the flight taking off from the airport in Las Vegas, he finds a slot machine to play. If he pays c dollars to pull the lever, he can win 1 dollar with probability q, and zero with probability $1 - q$. of course, he can decide not to play. Unfortunately, Matthew does not know q; however, he can summarize his belief on q by forming a probability distribution $f(q), q \in [0, 1]$, which can be updated by playing more times.

The one-armed bandit problem can be modeled by defining process 0 corresponding to do-not-play and process 1 corresponding to play. $S^0 = \{0\}$, $r^0 = 0$, and $p^0(0 \mid 0) = 1$. For the option of play, f is a probability density function defined on $[0, 1]$, which is also referred as the *prior distribution* in Bayesian reasoning. Let Q denote the radon variable of return in one play action, which has density f. The reward of one play is therefore

$$r^1(f) = \mathbb{E}_f[Q] - c \tag{4.52}$$

One way to model the Markov process in this example is due to the update knowledge about q. Suppose the revised (posterior) distribution f'. According to Bayes theorem, the transition probabilities satisfy

$$p^1(f \mid f') = \begin{cases} \mathbb{E}_f[Q], & f' = \dfrac{qf(q)}{\mathbb{E}_f[Q]} \\[3mm] 1 - \mathbb{E}_f[Q], & f' = \dfrac{(1-q)f(q)}{1 - \mathbb{E}_f[Q]} \end{cases} \tag{4.53}$$

Suppose the probability for Matthew to win in this round is $\mathbb{E}_f[Q]$. According to the Bayes theorem, we have

$$P(Q \approx q \mid win) = \frac{qf(q)}{\int_0^1 qf(q)dq} \tag{4.54}$$

which is computationally infeasible for general cases. We may adopt two possible approaches to resolve this dilemma:

- Choose a parametric family or *conjugate family* of densities that is closed for computations in (4.54).
- Represent the state space $\mathcal{S}^1$ corresponding to the number of wins and losses.

▶ **Exercise (Admission Control):** In the computer networks, the source node has packets to transmit through certain path resulting in a delay reaching the destination node. To avoid traffic jam inside the network, the mechanism of admission control is created as follows: At the source nodes, a queue of packets is formed. Once the source node gets the confirmation of a successfully received packet from the destination node, a permit of transmission is issued to transmit a packet in the queue. By this way, we can ensure the maximum number of packets in the network. Please model optimal admission control using a MDP.

4.3 Decision Making and Planning: Dynamic Programming

In the previous section, a general approach to solve the MDP and obtain an optimal policy is illustrated. As long as the probability distribution relating to the states and rewards are given, the best strategy can be easily derived. In fact, this is the typical scenario of control problems. Basically, the terms *prediction* and *control* relates to the task in an MDP:

- The environment's dynamics, the system dynamics, and the computation (or say, estimation) of the expected feedback from anything outside the decision maker is a prediction problem.
- On the basis of knowledge about the environment or the system, an optimal policy can provide the ideal control for decision maker.

Example: Tom, living in Tampa, is visiting San Francisco in three days. He has never been to a Lakers' game and carves to watch it once but he can only

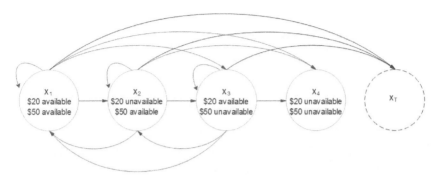

Figure 4.10　State-transition diagram.

afford at most $50 for the ticket. He checks TicketExchange.com, finding seats costing $50 are the only ones available. However, before the game starts it is always possible for the cheaper tickets to be released again. Except for the option of buying the currently available ticket first and refund it whenever he sees a cheaper one, when should he purchase the ticket so that he can watch the game without spending too much money?

Solution: Assume Tom checks out the website every hour, and the ticket's availability is the state. He draws down the state transition diagram. Noted in this case, the reward structure can be defined by ourselves. We let reward be 30 if Tom buys a $20 ticket, 0 if he buys a $50 ticket or does not purchase, and -20 if he doesn't get any ticket. It is a very straightforward way to model the reward by how much money he could save. As for the negative reward, imagine he needs to buy a $70 ticket when the cheaper ones are all sold out. Besides, whenever Tom purchases a ticket, state will transit into a terminal state, denoted by x_T. The value of terminal state must be zero recalling that the definition of state value is the expected future reward. Since there is no ongoing process after terminal state, the future reward is certainly zero. The state-transition of this Markovian system is shown as Figure 4.10.

Obviously, if any $20 ticket is available, Tom should immediately book it. Take state x_1 for example, he has three options a_1 (buy one for $20), a_2 (buy one for $50) and a_3 (buy nothing).

$$q(x_1, a_1) \doteq \sum_{x'} p(r, x'|x_1, a_1)[r + \beta v(x')] = 30 + \beta v(x_T) = 30$$

$$q(x_1, a_2) \doteq \sum_{x'} p(r, x'|x_1, a_2)[r + \beta v(x')] = 0 + \beta v(x_T) = 0$$

$$q(x_1, a_3) \doteq \sum_{x'} p(r, x'|x_1, a_3)[r + \beta v(x')]$$
$$= p_{11}[0 + \beta v(x_1)] + p_{12}[0 + \beta v(x_2)]$$
$$+ p_{13}[0 + \beta v(x_3)] + p_{14}[0 + \beta v(x_4)]$$
$$< 30$$

where p_{ij} represents the transition probability from x_i to x_j. The last inequality exists owing to the fact that the maximal state value is 30, with the discounted factor, $q(x_1, a_3)$ is definitely smaller than 30. Therefore, according to greedy policy, Tom shall select the action that maximize the action-value function–buying $20 ticket.

While current state is x_2, he has options a_1 (buy one for $50) and a_2 (buy nothing). The action values are

$$q(x_2, a_1) \doteq p(r, x'|x_2, a_1)[r + \beta v(x')] = 0$$
$$q(x_2, a_2) \doteq p(r, x'|x_2, a_2)[r + \beta v(x')]$$
$$= p_{21}[0 + \beta v(x_1)] + p_{22}[0 + \beta v(x_2)]$$
$$+ p_{23}[0 + \beta v(x_3)] + p_{24}[0 + v(x_4)]$$

As far as he knows, $v(x_1) = v(x_3) = 30$, $v(x_4) = -20$ whereas $v(x_3)$ remains unknown. But without any knowledge about how the ticket sales will change, he is not able to compare two action-value and hence can not make an optimal decision. Only if a genie would tell him the state transition probabilities, he can decide whether to buy the ticket now, which also explains the importance of the knowledge in the decision process.

Example (Cognitive Radio): Consider a set of frequency bands to represent the general case, though more dimensional radio resource can be considered. Suppose the frequency bands that we are interested in (typically PS operating) are a set of numbered bands, $\mathcal{M} = \{1, 2, \ldots, M\}$. At time t_n, cognitive radio (CR) operation allows an update of spectrum utilization. The nth observation (or allocation) time interval is $[t_n, t_{n+1})$. Due to opportunistic nature of each link (thus frequency band) modeled as a Markov chain, the ith frequency band is available following a Bernoulli process with probability π_i available, and is invariant to time.

We define the following indicator function, just as the clear channel indicator defined in carrier sense multiple access (CSMA) protocol in IEEE

802.11 wireless local area networks:

$$\mathbf{1}_i[n] = \begin{cases} 1, & \text{if channel } i \text{ is available in } [t_n, t_{n+1}) \\ 0, & \text{otherwise} \end{cases} \tag{4.55}$$

For perfect spectrum sensing, we can determine (4.55) in a reliable way. However, any spectrum sensing has some vulnerable situations, and thus we need to consider more to decide medium access control. Following (4.55), the probability mass function (pmf) of Bernoulli random variable at the ith frequency band is

$$f_{\mathbf{1}_i[n]}(x|\pi_i) = \pi_i x + (1 - \pi_i)\delta(x) \tag{4.56}$$

It is reasonable to assume $\{\mathbf{1}_i[n]\}_{i=1}^M$ independent, $n = 1, \ldots, L$ where L implies the observation interval depth. Denote $\boldsymbol{\pi} = [\pi_1, \ldots, \pi_M]$. For reliable CR operation, spectrum sensing is necessary, so that CR-Tx can have information about availability of each frequency band. However, for network operations on top of CR links, the strategy would be highly related to $\boldsymbol{\pi}$.

Case 1 $\boldsymbol{\pi}$ is known.

Case 2 $\boldsymbol{\pi}$ is unknown.

Case 3 $\boldsymbol{\pi}$ can be detected or estimated via some CR sensing or tomography methods [4].

Traditional CR functions as follows: At time t_n, CR learns the availability of a selected frequency band s_n (typically via spectrum sensing). If $\mathbf{1}_{s_n}(n) = 1$, information amount B can be successfully transmitted. For L time durations, the overall throughput is

$$W = \sum_{n=1}^{L} \mathbf{1}_{s_n}(n) \tag{4.57}$$

In case $\boldsymbol{\pi}$ is known, the spectrum sensing strategy for a CR is simply to select channel $i = \arg\max_{i \in \mathcal{M}} \pi_i$ to sense. Then, access decision is therefore optimally or suboptimally made based on certain decision criterion and conditions, while partially observed Markov decision process perfectly fits mathematical modeling for this situation.

So far, we considered just the scenarios relating to control problems, that is, we have assumed a perfect model of the environment's dynamics is known. Even though most of the tasks we face do not meet the assumption, a collection of algorithms called *dynamic programming* which is useful for computing optimal control/policy is still worth to be introduced. In fact, when the system model is unknown, the risen part of problems, namely prediction, can be dealt separately.

The underlying arguments that support dynamic programming are Proposition 2 and 3 from the previous section. Based on Proposition 2, an algorithm named *value iteration* promises to find out the optimal policy along with optimal value functions. Another approach *policy iteration*, rooting on Proposition 3, also leads a way to search for an optimal policy as well as optimal value functions.

Proposition 4 (Value Iteration Algorithms):

(i) Arbitrarily initialize $v(x)$ for all $x \in S$ (e.g. $v(x) = 0, \forall x \in S$)
(ii) For each state $x \in S$, assign $v(x)$ to $g(x)$
(iii) For each state $x \in S$, update its value function in respect of $g(x)$

$$v(x) = \max_{a \in A(x)} \sum_{x'} \sum_{r} p(r, x'|x, a) \left[r + \beta g(x') \right]$$

(iv) If $\sum_{x \in S} |g(x) - v(x)| < \Delta$ where Δ is a sufficiently small number, goes to step (v), else goes back to (ii)
(v) Output the optimal policy π^* such that

$$\pi^*(x) = \arg \max_{a \in A(x)} \sum_{x'} \sum_{r} p(r, x'|x, a) \left[r + \beta v(x') \right], \quad \forall x \in S$$

Proposition 5 (Policy Iteration Algorithms):

(i) Arbitrarily initialize $v(x)$ and $\pi(x) \in A(x)$ for all $x \in S$
(ii) For each state $x \in S$, assign $v(x)$ to $g(x)$
(iii) For each state $x \in S$, update its value function under the current policy π

$$v(x) = \sum_{x'} \sum_{r} p(r, x'|x, \pi(x)) \left[r + \beta v(x') \right]$$

(iv) If $\sum_{x \in S} |g(x) - v(x)| < \Delta$ where Δ is a sufficiently small number, goes to step (v), else goes back to (ii)
(v) For each state $x \in S$, assign $\pi(x)$ to $\delta(x)$
(vi) For each state $x \in S$, update its decision rule with regard to $v(x)$

$$\pi(x) = \arg \max_{a \in A(x)} \sum_{x'} \sum_{r} p(r, x'|x, a) \left[r + \beta v(x') \right], \quad \forall x \in S$$

(vii) If $\delta(x) = \pi(x)$ for every state $x \in \mathcal{S}$, goes to (viii), otherwise goes back to (ii)

(viii) Output the optimal policy $\pi^* = \pi$

From step (ii) to (iv) in policy iteration is the process of *policy evaluation*. Within these steps, we keep revising the state value under the same policy until the value functions converge. In the following steps (v) to (vii), based on the convergent value functions, we attempt to improve the current policy. This is so called *policy improvement*. Policy evaluation helps us understand how good the policy is and hence bring forth improvement in policy. On the other hand, value iteration algorithms do not try to maintain the policy. The only update that happens in step (iii) is related to the previous value function and has nothing to do with policy. The optimal policy is derived at the time when optimal state value comes out.

▶ **Exercise (Switching Job):** At the beginning, Geoffrey received a job offer to earn $1,000 per week. He may work to get weekly salary as the offer for the entire week, or seek an alternative employment to start next week (i.e. no income this week). If he decides to work in the current week, there exists 10% chance for him to lose the job next week (i.e. 90% chance to keep the job next week). If he seeks an alternative employment, there is no income in this current week, but the alternative weekly income will remain the same with probability 0.8, increase 5% with probability 0.12, or decrease 5% with probability 0.08.

 (a) Please generally formulate the problem as a MDP to maximize his total income over a finite horizon T, by defining q=10%, w=$1,000, ϵ=5%, $s_+ = 0.12$, and $s_- = 0.08$.

 (b) In case $T = 50$, numerically find his expected incomes.

 (c) Please identify the general strategy for Geoffrey to seek alternative or not in each week.

▶ **Exercise (Traveling Salesman Problem):** Traveling salesman problem (TSP) is the most well know dynamic programming problem. Suppose this salesman is going to n cities and we know the distance between city i and city j as $d(i, j)$, both $i, j \in \{1, 2, \ldots, n\}$1. Please find the shortest path to visit all cities. Direct computation results in an NP hard problem of complexity growing with n. Therefore, we need an algorithm to effectively compute this problem.

▶ **Exercise (Inventory Management):** A car dealer selling high-price sports cars for the brand L faces a weekly demand distribution as follows.

Number of Cars in Weekly Demand	0	1	2
Probability	0.3	0.5	0.2

The dealer looks at the end of each week and determines whether an order should be placed and subsequently the number of cars to order. The ordered cars arrive 1 week after the order being placed. Due to the capital of this dealer, there should be no more than 5 cars in total, either in inventory or on order. The cost of placing an order is $900, independent of the number to order. If there is a demand for a car but there is no inventory, sales profit suffers $3,500 loss. The cost (including interests and management etc.) associated with an unsold car in inventory is $300 per week. Please formulate this optimization as a Markov decision process and find the optimal policy.

4.4 Application of MDP to Search A Mobile Target

Suppose a mobile target operating by AI can move among X possible grids (or tiles). Since this mobile target can only move to geographical neighboring grids, its moving behavior can be modeled as a Markov chain whose transition probability matrix is $\mathbb{P}$. We intend to design a searching robot to select an action from the action space $\mathcal{A}$, at time instant $k \in \{0, 1, 2, \ldots, K\}$, where $K \le \infty$. An action $a \in \mathcal{A}$ can be a specific grid or a group of grids, such as the line-of-sight observation in Figure 4.11. At time k, an action a by the search robot can be executed with probability $1 - q(a)$, since the observation can be blocked with probability $q(a)$. Similar to radar detection, for any non-blocked grid(s) due to action a, there exists possibility of missing $\beta(a)$, that is, probability of detection is $1 - \beta(a)$. For the AI computing of a search robot knowing $\mathbb{P}, q, \beta$, how to search the grids to find the mobile target?

Suppose a search robot can act in $K < \infty$ consecutive time instants to find the mobile target, which suggests a finite-horizon MDP. Given the mobile target moving within X grids, denote the argument state space as $\mathcal{X} = \{1, 2, \ldots, X, T\}$, where T corresponds to a fictitious terminal state that means to terminate search if the target is detected prior to K searches. The subsequent observation space is denoted as $\mathcal{Y} = \{F, \bar{F}, B\}$, where F

Figure 4.11 If a search robot can conduct line-of-sight observation as the orange lines, with black grids as blocked and non-observable, the blue search robot can observe the red target, but the green search robot can not.

means "target found" and B means "search blocked". The search problem is formulated as follows:

- **Markov State Dynamics:** To model the search as a finite horizon MDP, the location of mobile target is treated as a finite-state Markov chain with transition probability matrix $\mathbb{P}$ by defining $x_k \in \mathcal{X}$ as the state of the mobile target at time instance $k = 0, 1, 2, \ldots, K$. To model the termination dynamics of search process after the mobile target is found, the observation transition probability matrices $\mathbb{P}^y$, $y \in \mathcal{Y}$:

$$\mathbb{P}^F = \begin{bmatrix} 0 & 0 & \cdots & 1 \\ \vdots & \vdots & \ddots & \vdots \\ 0 & 0 & \cdots & 1 \end{bmatrix} \tag{4.58}$$

and

$$\mathbb{P}^{\bar{F}} = \mathbb{P}^B = \begin{bmatrix} \mathbb{P} & \mathbf{0} \\ \mathbf{0}^T & 1 \end{bmatrix} \tag{4.59}$$

where $P(x_{k+1} = j \mid x_k = i, y_k = y) = \mathbb{P}^y_{ij}$ and the terminal state T is an absorbing state that occurs only when the mobile target is detected. The initial state is $\pi_0(i)$, $i \in \{1, 2, \ldots, X\}$.
- **Action:** The search robot selects action $a_k \in \mathcal{A}$, where $\mathcal{A}$ has actions to search one of the grids or a group of grids.

- **Observation:** At time instance k, $y_k \in \mathcal{Y} = \{F, \bar{F}, B\}$. Define the blocking probabilities $q(a)$ and the probabilities of missing $\beta(a)$. $\forall a \in \mathcal{A}, j = 1, 2, \ldots, X$,

$$P(y_k = F \mid x_k = j, a_k = a)$$
$$= \begin{cases} (1 - q(a))(1 - \beta(a)), & a \ searching \ grid \ j \\ 0, & otherwise \end{cases}$$
$$P(y_k = \bar{F} \mid x_k = j, a_k = a)$$
$$= \begin{cases} (1 - q(a)), & a \ not \ searching \ grid \ j \\ \beta(a)(1 - q(a)), & otherwise \end{cases}$$
$$P(y_k = B \mid x_k = j, a_k = a) = q(a) \tag{4.60}$$

For terminal state T, it is obvious that

$$P(y_k = F \mid x_k = T, a_k = a) = 1$$

- **Reward:** As introduced earlier, $r(x_k, a_k)$ denotes the reward for selecting a_k when the mobile target in state x_k. Reward design is an important incentive mechanism to tilt robot's actions reaching the purpose of control. There are several ways to define the reward function:
  (i) *Maximizing probability of detection:* The reward is the probability of detecting the mobile target (i.e. observation of F).

$$r(x_k = j, a_k = a) = P(Y_k = F \mid x_k = j, a_k = a), \ j = 1, \ldots, X \tag{4.61}$$
$$r(x_k = T, a_k = a) = 0 \tag{4.62}$$

(ii) *Minimizing search delay:* Any search step until reaching the terminal state T deducts one unit of reward (or generate one unit of cost).

$$r(x_k = j, a_k = a) = -1, \ j = 1, \ldots, X \tag{4.63}$$
$$r(x_k = T, a_k = a) = 0 \tag{4.64}$$

Each search step may consume different level of resources and thus equation (4.63) can be $c(a)$, cost associated with the action a.
- *Performance criterion:* Let $\mathcal{I}_k$ be the historical information available at time k. Then,

$$\mathcal{I}_0 = \pi_0; \quad \mathcal{I}_k = \{\pi_0, a_0, y_0, \ldots, a_{k-1}, y_{k-1}\}, k = 1, \ldots, K \tag{4.65}$$

A *search policy* π is a sequence of decision rules $\pi = \{\delta_0, \ldots, \delta_{K-1}\}$ where $\delta_k : \mathcal{I}_k \to \mathcal{A}$. The performance criterion is consequently the job function

$$J_\pi(\pi_0) = \mathbb{E}_\pi\{\sum_{k=1}^{K-1} r(x_k, \delta_k(\mathcal{I}_k)) \mid \pi_0\} \tag{4.66}$$

The *optimal search* is now to find the policy to maximize (4.66) for all initial distributions, that is,

$$\pi^* = \underset{\delta \in \mathcal{A}}{\operatorname{argmax}} J_\pi(\pi_0) \; \forall \pi_0 \in \Pi(X) \tag{4.67}$$

Remark: Instead of the reward function $r(x_k, a_k)$, the cost function $c(x_k, a_k)$ is also widely used in MDP, which has opposite meaning to reward but is equivalent in optimization by changing maximization for reward to minimizing for cost. The intuitive relationship between reward and cost can be $c(x_k, a_k) = -r(x_k, a_k)$ or $c(x_k, a_k) = 1/r(x_k, a_k)$. If the cost function in (4.66) is used, then we consider to minimize in (4.67).

▶ **Exercise:** The red star represents for a mobile target randomly either moving forward or turning right/left in the white grids of Figure 4.12, where "randomly" means equally probable among possible action set of $\{forward, left, right\}$. Without knowing the location of red star, a search algorithm shall be developed based on MDP including (i) selecting the starting grid (ii) continuing search for one of neighboring grids (up, down, left, right) and the same grid (iii) probability of missing is $P_M = 0.01$ and probability of blocking is $q(\circ) = 0.6$ for repeated observation in the same grid. Since the red star is prohibited to black grids, any black grid is not

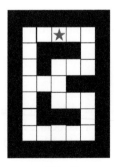

Figure 4.12   The mobile target moves within the area of grids while black grids means blocking.

Figure 4.13 Exploration versus exploitation.

considered into red star's movement and thus in the search process. Please find the optimal MDP search in terms of search time (i.e. steps).

4.5 Multi-Armed Bandit Problem

MDP actually reveals a dilemma, exploration vs. exploitation. Nathan usually goes to a restaurant for lunch and knows pretty well about what he can get for lunch with satisfaction. However, there is a new restaurant just opening, which provides a possible chance for better food but may experience less tasty food. Shall Nathan go to the original one or the new one? This kind of situations are common for decisions by smart agents.

If we have learned all the information about the environment, we are able to find the best strategy by just simulating brute-force or many other intelligent approaches. The dilemma comes from the incomplete information: we need to gather enough information to make best overall decisions while keeping the risk under control. With exploitation, we take advantage of the best option we know. With exploration, we take some risk to collect information about unknown options. The best long-term strategy may involve short-term sacrifices. For example, one exploration trial could be a total failure, but it warns us of not taking that action too often in the future. Such tradeoff is generally useful in robotics and AI.

Now, let us examine an interesting problem. If one is over 21 years old, once one gets into a casino and sees a line of slot machines as Figure 4.14, then what is the optimal strategy to win by playing with these machines? This problem is known as *multi-armed bandit* (MAB) problem. MAB was first introduced by Robbins in 1952, and has since been used extensively to model the trade-offs faced by an automated agent who aims to gain new knowledge

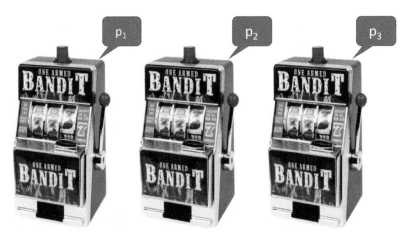

Figure 4.14 Multi-armed bandit problem, with unknown probability to win.

by exploring its environment and to simultaneously exploit its current and reliable knowledge. MAB actually has rich insights and can be applied to many diverse problems. For example,

- Clinical trials: In traditional clinical trials, patients are randomized into two equal-sized groups. The better treatment can usually be identified with a high level of confidence, with half of patients in testing. Adaptive trials dynamically allocate more patients to the better treatment. Modern adaptive clinical trials can be classified into K families. Group sequential trials are designed based on the way that trials can be stopped prematurely based on interim results, such as the performance of a particular treatment. Sample size re-estimation allows designs of the patient population size to be readjusted in the trial. Drop-the-losers designs, in particular, allow certain treatments to be dropped or added. Naturally, such trials drop less promising treatments first [6].
- Routing in networks: A typical routing problem in a communication network is to select a networking path among K candidate paths, while an equivalent problem is to assign a job to one of the parallel or distributed processors. Each path is associated with its bandwidth/capacity and delay. Consequently, each path can be viewed as an arm of MAB.
- Online advertising: Each time a user visits a website, one of the K possible advertisements will be displayed. A reward is granted if a user clicks on the advertisement. No *prior* knowledge of the user, advertisement content, webpage content, etc. is required.

- Economic applications: In addition to the gambling problem, a lot of financial scenarios can be applied, and further problems like the computation of Nash equilibria in game theoretic applications are also equivalent to MAB.

The MAB problem well demonstrates the dilemma of exploration versus exploitation. Back to the problem in Figure 4.14, an naive approach is to keep playing with one slot machine for many rounds (i.e. episodes in learning or a dynamic system). Based on the law of large number, we can eventually obtain the "true" probability for this slot machine. However, this way can be quite wasteful in resource and no guarantee in finding global optimal reward.

Given greedy policy (i.e. to keep playing slot machines), one possible theoretic treatment is to maximize the reward obtained by successively playing these slot machines (i.e. the arms of the bandits). The MAB problem in Figure 4.14 can be treated as a *Bernoulli multi-armed bandit* described as a tuple of $(\mathcal{A}, \mathcal{R})$ of action space and reward, as follows:

- There are K slot machines with reward probability $p_1, p_2, \ldots, p_K$.
- At each time instance or episode, the agent takes an action $a_t \in \mathcal{A}$ and receives an immediate reward r_t.
- $\mathcal{A}$ is the action space, corresponding to the interaction with a slot machine. The return of action a is the expected reward, $Q(a) = \mathbb{E}[r \mid a] = p$. Specifically, action a_t on slot machine i at time t implies $Q(a_t) = p_i$.
- $\mathcal{R}$ is a reward function, which is stochastic for a Bernoulli bandit. At time t, $r_t = \mathcal{R}(a_t)$ returns reward "1" with probability $Q(a_t)$ or "0" with probability $1 - Q(a_t)$.

We may note that this is precisely a simplified version of Markov decision process without the state space $\mathcal{S}$. The goal is to maximize the cumulative reward R_T within time horizon T as

$$R_T = \sum_{t=1}^{T} r_t \tag{4.68}$$

The optimal reward probabilities $p*$ of the optimal action $a*$ is

$$p* = Q(a*) = \max_{a \in \mathcal{A}} Q(a) = \max_{1 \leq i \leq K} p_i \tag{4.69}$$

If we know the optimal action with the best reward, the goal is equivalent to minimizing the potential *regret* or loss by not selecting the optimal action.

The total loss or the total regret is therefore

$$\mathcal{L}_T = \mathbb{E}\left[\sum_{t=1}^{T}(p*-Q(a_t))\right] \tag{4.70}$$

Earlier argument shows naive or pure exploitation not good enough. A good bandit strategy shall incorporate exploration one way or another. A few widely adopted algorithms are oriented in the following.

▶ **Exercise (Switching Job):** Please solve this problem in Section 4.3 using MAB.

4.5.1 ϵ-Greedy Algorithm

We may consider *greedy algorithms* as the simplest and most common approach to online decision problems, which involve two steps to generate each action: (1) estimate a model from historical data and (2) select the action that is optimal for the estimated model, and arbitrary for any tie. Such algorithms are greedy in the sense that an action is chosen solely to maximize immediate reward.

Similar to the evolutionary algorithm of small mutation probability, the ϵ-greedy algorithm exploits the most promising actions for most of the time (with probability $1 - \epsilon$), but conducts random exploration occasionally (with probability ϵ). The action value is estimated according to the past experience and is implemented by averaging the rewards associated with the observed target action $a_{1:t}$. In Bernoulli Bandit as Figure 4.14, we have

$$\hat{Q}_t(a) = \frac{1}{N_t(a)} \sum_{\tau=1}^{t} r_t \cdot \mathbb{I}_{a_\tau = a} \tag{4.71}$$

where $\mathbb{I}$ is an indicator function and $N_t(a) = \sum_{\tau=1}^{t} \mathbb{I}_{a_\tau = a}$ is the counting variable for the number of times that a specific action a has been selected. For most of the time (i.e. with probability $1 - \epsilon$), the agent exploits the best possible action that have learned up to time t, $a_t^* = \mathrm{argmax}_{a \in \mathcal{A}} \hat{Q}_t(a)$, while proceeds random exploration with a small probability ϵ.

4.5.2 Upper Confidence Bounds

Random explorations give opportunities to try options that the agent has not known much. However, the randomness may lead to a bad action that has been already known. To avoid such inefficient explorations, we may conduct

(a) to decrease the parameter ϵ in time

(b) to be optimistic about those options of high uncertainty and thus to prefer actions that have not got confident value estimations yet, which suggests that the agent should explore actions of higher potential reaching optimal value(s).

The *upper confidence bounds* (UCB) algorithm aims at supplying a measure of such potential by an upper confidence bound of the reward value, $\hat{U}_t(a)$, such that the true value $Q(a)$ is below the bound $\hat{Q}_t(a) + \hat{U}_t(a)$ with high probability. This upper bound $\hat{U}_t(a)$ shall be a function of $N_t(a)$, and a larger number of trials $N_t(a)$ suggests a tighter bound.

Applying the UCB algorithm, the agent typically selects the greediest action to maximize the upper confidence bound.

$$a_t^{UCB} = \underset{a \in \mathcal{A}}{\operatorname{argmax}} \, \hat{Q}_t(a) + \hat{U}_t(a) \tag{4.72}$$

The remaining question is how to estimate this upper confidence bound. In case there is no prior knowledge on distribution, *Hoeffding's Inequality* applicable to any bounded distribution is useful. Let $X_1, \ldots, X_t$ be independent and identically distributed (i.i.d.) random variables bounded by the interval $[0, 1]$. The *sample mean* is thus $\bar{X}_t = \frac{1}{t} \sum_{\tau=1}^{t} X_\tau$. For $u > 0$,

$$\mathbb{P}[\mathbb{E}(X) > \bar{X}_t + u] \leq e^{-2tu^2} \tag{4.73}$$

For a specific action $a \in \mathcal{A}$, then we can view

- $r_t(a)$ as a random variable
- $Q(a)$ as the true mean
- $\hat{Q}_t(a)$ as the sample mean
- $u = \hat{U}_t(a)$ as the upper confidence bound

Consequently,

$$\mathbb{P}\left[Q(a) > \hat{Q}_t(a) + U_t(a)\right] \leq e^{-2tU_t^2(a)} \tag{4.74}$$

In other words, it is of high probability that the true mean is below the sum of sample mean and upper confidence bound. Since $e^{-2tU_t^2(a)}$ is small, we define

$$\rho = e^{-2tU_t^2(a)} \tag{4.75}$$

Then, to complete UCB algorithm,

$$U_t(a) = \sqrt{\frac{-\log \rho}{2N_t(a)}} \tag{4.76}$$

It is desirable to reduce ρ in time. To develop more accurate estimation of confidence bound with more observations, by setting $\rho = t^{-4}$, we can obtain the *UCB1* algorithm as follows.

$$U_t(a) = \sqrt{\frac{2 \log t}{N_t(a)}} \tag{4.77}$$

$$a_t^{UCB1} = \underset{a \in \mathcal{A}}{\mathrm{argmax}}\, Q(a) + \sqrt{\frac{2 \log t}{N_t(a)}} \tag{4.78}$$

For UCB and UCB1 algorithms, no *prior* knowledge of reward distribution is assumed, and Hoeffding's inequality is therefore employed to generally give estimation. However, in many cases, it is possible to obtain some prior information to develop *Bayesian UCB*. For example, in the multi-armed bandit problem, if the reward of each slot machine is Gaussian, we would be able to establish the 95% confidence interval to set $\hat{U}_t(a)$.

4.5.3 Thompson Sampling

Thompson sampling is a rather simple idea but surprisingly works well.

Example (Probability Matching): In a binary decision problem, during the training period, H_1 is true with 80% chances and H_0 holds with 20% chances. What is the optimal Bayesian strategy to maximize the number of correct predictions? It is straightforward to selection predictive actions by probability matching, which suggests $(0.8)^2 + (0.2)^2 = 0.68$ probability to correctly predict. However, a naive decision mechanism to always predict H_1 gives 80% probability of correct prediction. This example of probability matching turns out useful.

Back to MAB problem. In each time episode, an action $a \in \mathcal{A}$ is selected according to the probability that a is optimal

$$p^*(a \mid h_{0:t}) = \mathbb{P}\left[Q(a) > Q(a'), \forall a' \neq a \mid h_{0:t}\right] \tag{4.79}$$

$$= \mathbb{E}_{\mathcal{R}\mid h_{0:t}}\left[\mathbb{I}_{\mathrm{argmax}_{a \in \mathcal{A}} Q(a)}\right] \tag{4.80}$$

where $p^*(a \mid h_{0:t})$ is the probability of taking action a given the historical data.

Example (Bernoulli Bandit): For a K-armed Bernoulli bandit, $a_t = k \in \mathcal{A} = \{1, 2, \ldots, K\}$ yields a success with probability $p_k \in [0, 1]$. The success probabilities $p_1, \ldots, p_K$ are unknown to the player (i.e. agent) and fixed in time, which can be learned by experimentation. The objective is to maximize R_T defined earlier, where T is usually much larger than K. A naive approach to this Bernoulli bandit problem involves allocating some fixed fraction of time periods to exploration by the manner that an arm is uniformly sampled at random, while aiming to select successful actions in other time periods. This precisely seeks best strategy between exploration and exploitation. However, after well studied in decision science and control engineering, such an approach can be still wasteful even for such simple Bernoulli bandit. For application of Bernoulli bandit to online advertisement, the arms correspond to the different banner advertisements that can be displayed on a website. A success corresponds to a click on the advertisement, while p_k represent either the click-through-rate or conversion-rate among the population of users who visit the website.

The more realistic and useful application scenarios are not only to learn from historic data, but also to explore systematically to improve future performance. For example, Google maps or Apple maps have a function that the shortest (either distance or traveling time) path(s) are recommended once you select the source and destination based on the real-time traffic conditions, while the blue route is recommended with two grey alternatives in Figure 4.15. A simplified version of this online shortest path can be given as the following example.

Example (Online Shortest Path Problem): Christine drives from home (K-Bar Ranch) to her office at the University of South Florida (USF) in the morning. She would like to commute along the path that takes the least expected travel time, but she is uncertain of the travel time along different routes. How can she efficiently learn and minimize the total travel time after a large number of trips?

Solution: We form this online shortest path problem by creating a graph $\mathcal{G} = (\mathcal{V}, \mathcal{E})$ as Figure 4.16, where the source vertex is Lake Hanna and the destination vertex is USF. Each vertex can be thought of as an intersection, and for two vertices $i, j \in \mathcal{V}$, an edge $(i, j) \in \mathcal{E}$ is present if there exists a direct road segment connecting the two intersections. Suppose that traveling along an edge $e \in \mathcal{E}$ takes time duration θ_e on average. If these parameters were known, Christine would select a path $(e_1, \ldots, e_n)$, consisting of a sequence of adjacent edges connecting vertices 1 and $N = 10$, such that the expected total time $\theta_{e_1} + \cdots + \theta_{e_n}$ is minimized, as the conventional shortest path problem.

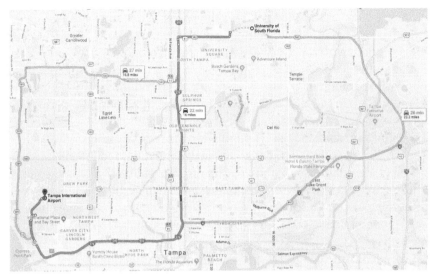

Figure 4.15 Online routing example: from university of south florida to tampa international airport.

For the online shortest path problem, Christine selects paths according to a sequence of time periods. In time period t, the realized time $\tau_{t,e}$ to traverse edge e is drawn independently from an empirical distribution with mean θ_e. Christine (i.e. the agent) sequentially selects a path x_t, observes the realized travel time $\tau_{t,e|e\in x_t}$ along each edge in the path, and incurs cost $c_t = \sum_{e\in x_t} \tau_{t,e}$ equal to the total travel time. By exploring intelligently, she wishes to minimize cumulative travel time $\sum_{t=1}^{T} c_t$ over a large number of periods T. Conceptually, this is similar to the Bernoulli bandit problem, but computationally infeasible. An efficient approach is therefore required to take advantage of statistical and computational structure of the problem.

The natural thinking suggests that the agent sets up a timer when leaving the source and check once arriving the destination, effectively only track the travel time of the selected path, which is the basic idea behind the Thompson sampling. This is close to the Bernoulli bandit model, where only the realized reward (or cost) of the selected arm would be observed. We may take further correlation among neighboring road segments into account.

As a conclusion, for many cases regarding MDP and MAB in robotics, the purpose is to develop the algorithm striking efficiency between exploration and exploitation.

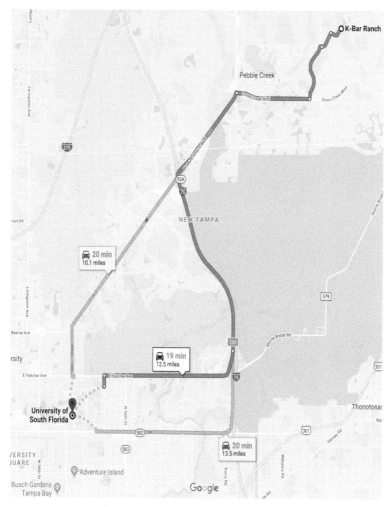

Figure 4.16 From K-Bar Ranch to USF.

■ **Exercise (Online Advertisement):** MAB is widely applied to online advertisement that is critical for e-commerce. This simple exercise is to demonstrate how to apply this knowledge. A website charges each online advertisement in the following way: $10 for 30 days and $3.50 per hit once beyond 10 hits in 30 days, while 30 days form a charge cycle to set from zero. According to many 30-day cycles, the total number of hits in 30 days is rather stable. Since 30 days ago, the website has published new advertisements A-H

Day	A	B	C	D	E	F	G	H
1	x	x			x			
2			x		x			x
3	x							xx
4				x			xx	
5		x						xx
6	x					x		
7			x				x	
8	x							x
9		x					x	
10	x							x
11					x	x		
12		x			x	x		
13		x		x			x	
14			x				x	
15	x			x				
16			x			x		
17	x						x	
18	x			x				
19		x			x			
20	x		xx					
21				x		x		
22					x	x		
23	x				x			
24		x					x	
25	x		x					
26			x			x		
27	x				x			
28		x						x
29	x			xx				
30				x		x		

Figure 4.17 Hit record for advertisements A-H.

with hit record as Figure 4.17 denoting a hit by a cross. For a new coming 30-day cycle, all the eight advertisements A-H want to continue. The CEO Sergio has a different view to look at the situation, aiming to cut down the number of online advertisements on the website. Do you agree Sergio's opinion? If not, what is the quantitative reason to reach this conclusion? If so, which advertisements to keep with your quantitative analysis?

■ **Exercise (Online Route Selection):** Julie drives from her home at Cheval to USF every day, where the map is provided in Figure 4.18. Suppose the traffic in morning rush hours is heavy and dynamic due to possible jamming

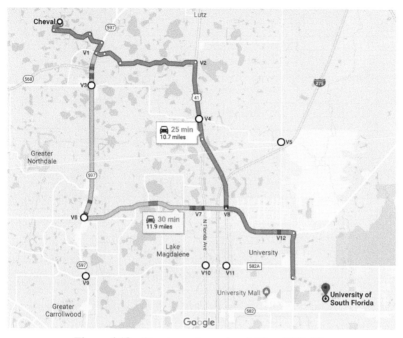

Figure 4.18 Street map from cheval (C) to USF (U).

and accidents. We intend develop the online software for Julie to drive in a time-efficient way.

(a) Suppose Julie leaves home at time T. Based on the the traffic situation, Figure 4.19, the software notifies Julie the best route to drive. What is this best route and the algorithm to identify it?

(b) Every 5 minutes, the traffic situations are updated and thus the driving time for each road segment is updated. Please develop the online algorithm and consequent software for online update routing. If the update route is different from the current route, Julie will get the notice to change and Julie follows the update route. What is the final route for Julie to arrive USF (please specify each change of routing)?

(c) What is the computational complexity for the entire trip in (b)?

(d) Figure 4.18 shows two most common routes for Julie according to long-term historical data. Please develop a more computationally efficient online routing algorithm. Compared with the algorithm in (b), is there difference in online route update and associated complexity and required memory?

	T	T+5	T+10	T+15	T+20	T+25	T+30	T+35	T+40
C-V1	3	3	3	3	3	3	3	4	5
V1-V2	6	6	6	6	8	8	8	6	5
V1-V3	2	3	4	2	2	2	1	1	1
V2-V4	2	2	3	2	2	2	2	2	2
V3-V4	8	8	8	8	8	8	9	9	8
V3-V6	5	8	11	15	12	9	7	5	5
V4-V5	5	4	3	4	6	5	5	4	5
V4-V7	4	4	3	3	4	5	4	5	4
V4-V8	4	5	4	4	5	4	4	4	4
V5-V12	5	3	4	3	3	8	6	5	4
V6-V7	8	9	9	8	8	8	9	12	11
V6-V9	3	3	4	3	3	3	4	2	3
V7-V8	2	2	1	1	3	2	2	2	2
V7-V10	3	3	3	3	3	3	4	3	3
V8-V12	5	9	7	5	5	6	5	5	6
V8-V11	3	3	3	4	3	3	3	3	3
V9-V10	9	10	11	11	10	9	10	9	10
V10-V11	2	2	2	1	3	2	2	2	3
V11-U	5	5	4	4	5	7	6	5	5
V12-U	5	9	7	5	4	5	5	6	6

Figure 4.19   Traveling time for each road segment with updates in every 5 minutes.

Remark: This exercise pretty much describes one of the principles for GPS route guidance.

Appendix: Markov Chains

A random process X_t is a collection of random variables indexed by (time) t. If t is countable, X_t is a *discrete-time* random process. Furthermore, a random process X_t is is a *markov process* if the future of the process given the present is independent of the past. That is,

$$P(X_{t+1} = x_{t+1} \mid X_t = x_t, \dots, X_1 = x_1) = P(X_{t+1} = x_{t+1} \mid X_t = x_t)$$
(4.81)

A discrete-valued Markov random process is a *Markov chain*. If the number of discrete values is finite, it is a *finite-state Markov chain* (FSMC), which is the primary focus in AI of robotics. The collection of state in a FSMC is denoted as $\mathcal{S}$. Let $\mathcal{S} = \{1, 2, \dots, N\}$, that is, there are N states in this FSMC. The state-transition probability is defined as

$$P_{ij}^t = P(X_{t+1} = j \mid X_t = i)$$
(4.82)

We are particularly interested in the homogeneous FSMC, which means that the the state-transition probabilities are invariant with time index. $\forall t$,

$$P_{ij} = P(X_{t+1} = j \mid X_t = i)$$
(4.83)

It is common to use *state transition diagram* to represent the behavior of a Markov chain, particularly a homogeneous FSMC. The state transition matrix is therefore defined as $\mathbb{P} = [P_{ij}]$, an $N \times N$ matrix.

If loop states and sink states do not exist in a FSMC, as time progress, FSMC will reach its steady-state. The probability distribution of steady-state for state n in the FSMC can be denoted by π_n^s, which can be computed by

$$(\pi_1^s, \ldots, \pi_N^s) \cdot \mathbb{P} = (\pi_1^s, \ldots, \pi_N^s) \tag{4.84}$$

$$\sum_{n=1}^{N} \pi_n^s = 1 \tag{4.85}$$

where above two equations are based on the invariant nature of steady-state and total probability.

Example: A robot works in an automated production line by executing the same task every minute. Its output is rated as "accepted" or "unaccepted" due to precision. The Markov property for robot's task is observed: If "accepted" one minute ago, then "accepted" with probability 0.9 (i.e. "unaccepted" with probability 0.1) in this minute. If "unaccepted" one minute ago, then "accepted" with probability 0.8 (i.e. "unaccepted" with probability 0.2) in this minute. The performance of this robot can be represented by a FSMC shown in Figure 4.20.

▶ **Exercise:** In a long run, what is the probability that the output of this robot is accepted in precision?

▶ **Exercise:** A robot randomly walks on a one-dimensional axis. Each time, the robot moves either to left or right by one step with equal probability (i.e. $1/2$). Suppose the robot starts from the origin of the axis.

(a) Once the robot moves to N_{win} or $-N_{loss}$, the process terminates. What is the probability to terminate at N_{win}? This is known as Gambler ruin problem, since N_{win} is usually large and N_{loss} is usually a relatively

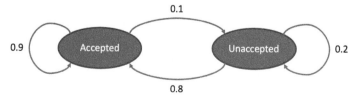

Figure 4.20 Two-state Markov chain.

small number (i.e. analogue to the capital of the player in a casino is usually a relatively small number), and thus the probability to terminate at N_{win} approaching 1 as N_{win} going to ∞.

(b) Suppose the robot moves left or right in different probabilities, say moving left has slightly less probability $0 < p < 1/2$. Once the robot reaches N_ν or $-N_\nu$, the process terminates. What is the probability to terminate at $-N_\nu$?

(c) What is the probability for this robot finally back to the origin once leaving it? Hint: This problem deals with an infinite-state Markov chain and please consult a book in Stochastic Processes such as [8].

Further Reading: [2] presents in-depth knowledge in signal detection and estimation. [3] supplies mathematical foundation of Markov decision processes. [5] supplies mathematical foundation of dynamic programming. More details about Thompson sampling can be found in [7] with Python programming.

References

[1] J.O. Berger, *Statistical Decision Theory*, 1985.

[2] H.V. Poor, *Introduction to Signal Detection and Estimation*, 2nd edition, Springer.

[3] M.L. Puterman, Markov Decision Processes, Wiley, 1994.

[4] C.-K. Yu, S.M. Cheng, K.-C. Chen, "Cognitive Radio Network Tomography", the special issue of Achievements and Road Ahead: The First Decade of Cognitive Radio *IEEE Transactions on Vehicular Technology*, vol. 59, no. 4, pp. 1980-1997, April, 2010.

[5] D.P. Bertsekas, *Dynamic Programming*, Prentice-Hall, 1987.

[6] V. Kuleshov, D. Precup, "Algorithms for the Multi-Armed Bandit Problems", *Journal of Machine Learning*, vol. 1, 2000.

[7] Daniel J. Russo, B. Van Roy, A. Kazerouni, I. Osband, Z. Wen, "A Tutorial on Thompson Sampling", arXiv: 1707.02038v2, 2017.

[8] S. Ross, *Stochastic Processes*, 2nd edition, Wiley, 1995.

5

Reinforcement Learning

In earlier chapter, we introduce a lot of typical applications of Markov decision process (MDP) including

- Inventory problem
- Routing
- Admission control
- Sequential resource allocation
- Secretary problem (i.e. dynamic programming)

When state-transition statistics and thus modeling of the dynamic system are known, MDP can lead to mathematically feasible cases, though still possible to suffer from computational complexity. However, we are more interested in the circumstances that deal the environment's dynamics, namely $p(r, x'|x, a)$ is not *a priori* known, or system model is not available. A branch of artificial intelligence, known as *reinforcement learning* (RL), is heavily related to the MDP and is very useful for addressing problems in this scenario. The essence of reinforcement learning is to learn from experience. A number of approaches can be normally categorized into:

- Model-based reinforcement learning: based on experience to learn $P(r, x'|x, a)$ and derive a policy consequently.
- Model-free reinforcement learning: based on experience to directly learn value function or evaluate different policies to search in the space of policy such as using gradient ascent search. For many robotic and AI problems, a precise model unlikely exists in practice, and model-free is a strong motivation to adopt machine learning techniques.

As a quick conclusion, the supervised learning deals with the inputs of data and label; the unsupervised learning deals with the inputs of data without label; reinforcement learning deals with the interaction with the environment, of particular interest to robotic problems.

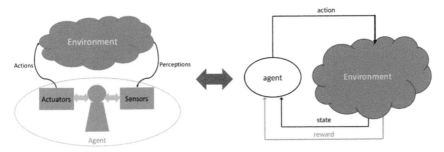

Figure 5.1   Robot and model of reinforcement learning.

Example: Playing Atari by deep Q-learning (a kind of reinforcement learning) is a good example. Please refer a Youtube video by Google DeepMind: https://www.youtube.com/watch?v=V1eYniJ0Rnk

5.1 Fundamentals of Reinforcement Learning

Three essential elements that lie in the reinforcement learning problem is respectively agent, reward, and environment. The agent interacts with the environment and equips with sensors to determine the state of the environment, in order to take an action to modify the state. When the agent takes an action, the environment provides a reward. It can be summarized as the following Figure. The agent intends to learn the best sequence of actions that result in the maximum cumulative rewards.

Among all the actions in the action set, which gives back higher reward is unsure. Only after continuous trials can the decision maker give qualitative estimation. Nonetheless, while carrying out decisions over epochs, the decision maker will face a trade-off between selecting experienced profitable actions (i.e. exploitation) or executing untried actions (i.e. exploration). The following sub-section presents a special case of MDP. which has been applied to many problems in computer systems, network systems, and management science, explains the importance of both exploitation and exploration.

5.1.1 Revisit of Multi-Armed Bandit Problem

This K-armed bandit is a hypothetical slot machine with K levers, but has only one state. The action is to select and pull one of the levers, while a certain reward is associated with such action. The goal is simply to decide which lever to pull to maximize the reward. This is a simple model due to

- Only one state (one slot machine)
- Only need to decide the action
- immediate reward (to observe the value from the action)

$V(a)$ is the value of action a. Initially $V(a) = 0 \; \forall a \in \mathcal{A}$. When we execute action a, we get the reward $r_a \geq 0$. If the reward is deterministic, $V(a) = r_a$. If the rewards are stochastic, the amount of rewards is defined by the probability distribution $p(r|a)$. We can define $V_t(a)$ as the estimate of the value from action a at time instant t, which can be the average of all rewards when action a was selected prior to time t.

Assume there are K bandit machines in the casino. Each machine generates rewards following normal distribution with distinct means μ_k. That is, if the gambler plays the kth machine, he gets rewards r_k where

$$p(r_k|a_k) = \frac{1}{\sqrt{2\pi}} \exp\left(-\frac{(r_k - \mu_k)^2}{2\pi}\right).$$

Besides, their means are reset at the beginning of day. John and Bob go to the casino every day, spending hours on gambling. John is conservative and provident. He always plays the machine that provides the highest average reward, that is, he selects the action (of playing certain bandit machine) with the maximum value. This *greedy policy* indicates John executes the optimal action

$$a^* = \arg\max_a V_t(a)$$

at all times. Bob is more adventurous, sometimes he does not want to make choices that are seemingly superior. Therefore, he plays the machine of the highest value with probability $1 - \epsilon$. The other three machines are played with equal chances. Following what we call ϵ-*greedy policy* ($\epsilon < 1$), Bob's action at epoch t is

$$a_t = \begin{cases} a^* & \text{with probability } 1 - \epsilon \\ a \neq a^* & \text{with probability } \frac{\epsilon}{|\mathcal{A}|-1} \end{cases} \tag{5.1}$$

Example: Given $K = 10$ and continuing daily 1000 plays for 2000 days, John and Bob average their first, second, third play and so on happened in these two thousand days.

They observe that John frequently locked onto the suboptimal action since the early stage (before roughly first 50 plays). Bob never gave up trying all the options, resulting in higher rewards at the end of each day averagely–although at the very beginning, he seems to earn less rewards than John.

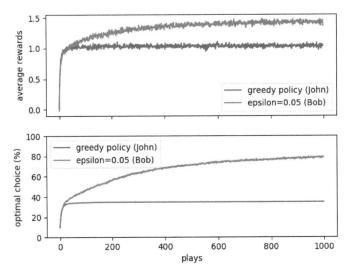

Figure 5.2   Comparisons of K-armed bandit policies, $K = 10$.

Remark: This example tells how important to involve both exploitation and exploration into a policy: exploitation to make the best possible decision based on current information, and exploration to gather more information. Nevertheless, the best long-term policy may result in short-term sacrifices.

When we want to exploit in the decision process, we select the action with the maximum value. That is,

$$a^* = \operatorname*{argmax}_{a} V(a) \tag{5.2}$$

Possible generalizations toward *reinforcement learning* include:

- More states, say different slot machines with different reward probabilities $p(r \mid x_i, a_j)$ and to learn $V(x_i, a_j)$, the value of taking action a_j in state x_i. Please note that we will use $Q(x_i, a_j)$ to denote state-action value $V(x_i, a_j)$ in the following text regarding reinforcement learning. Consequently, $Q(\cdot, \cdot)$ or Q-values specifically correspond to state-actions values.
- Actions affect not only the reward but also the next state.
- Rewards are delayed and we have to immediately estimate the values from the delayed rewards.

Looking closely for estimating the values of actions in K-armed bandit problem, an intuitive estimation can be obtained by averaging the rewards actually received:

$$Q_t(a) = \frac{\sum_{i=1}^{t-1} R_i \cdot \mathbb{I}_{A_i=a}}{\sum_{i=1}^{t-1} \mathbb{I}_{A_i=a}} \tag{5.3}$$

where A_i is the decision of K-MAB at time i, with reward R_i. The law of large number suggests $Q_t(a) \to q^*(a)$. The greedy selection of action implies

$$A_t = \underset{a}{\operatorname{argmax}}\, Q_t(a) \tag{5.4}$$

Q_{n+1} denotes the estimate of an agent's action value after n actions. With the following algebraic manipulations,

$$Q_{n+1} = \frac{1}{n} \sum_{i=1}^{n} R_i$$

$$= \frac{1}{n} \left(R_n + \sum_{i=1}^{n-1} R_i \right)$$

$$= \frac{1}{n} \left(R_n + (n-1)\frac{1}{n-1}\sum_{i=1}^{n-1} R_i \right)$$

$$= \frac{1}{n} \left(R_n + (n-1)Q_n \right)$$

$$= Q_n + \frac{1}{n} [R_n - Q_n]$$

It suggests an iterative format as

$$NewEstimate \leftarrow OldEstimate + StepSize\, [Target - OldEstimate] \tag{5.5}$$

which is useful to develop online algorithms. We therefore conclude online update as follows.

Proposition 1 (Online Update).

$$Q_{t+1}(a) \leftarrow Q_t(a) + \eta\, [r_{t+1}(a) - Q_t(a)] \tag{5.6}$$

where $r_{t+1}(a)$ is the reward after taking action a at time $t+1$; η is the learning factor (which can be gradually decreasing in time for the purpose of convergence); r_{t+1} is the desired reward output; $Q_t(a)$ represents the current prediction; $Q_{t+1}(a)$ is the expected value of action a at time $t+1$ converging to the mean of $p(r \mid a)$ as t increasing.

While facing a non-stationary environment, we may want to weight most recent rewards much more than long-past ones. For example, introducing a constant step size parameter $\alpha \in [0, 1]$,

$$Q_{n+1} = Q_n + \alpha [R_n - Q_n] \tag{5.7}$$

After iterations,

$$Q_{n+1} = (1 - \alpha)^n Q_1 + \sum_{i=1}^{n} \alpha (1 - \alpha)^{n-i} R_i \tag{5.8}$$

which suggests that the weight decays exponentially according to the exponent $1 - \alpha$, as an *exponential, recently-weighted average*.

5.1.2 Basics in Reinforcement Learning

The agent who is the decision maker in a learning process interacts with the environment and equips with sensors to determine the state of the environment, in order to take an action to modify the state. When the agent takes an action, the environment provides a reward. As time progressing, the agent develops sequential decision making to outside world. Up to now, the following classes of sequential decision making process have been identified:

Programming: An intelligent agent can be programmed to handle all possible situations. For each possible state, an action can be specified *a priori*, such as a convolutional decoder. However, due to the complexity or the uncertainty of the system, programming might not be feasible.

Search and Planning: Back to late 1990's, *Deep Blue* adopted brute force search algorithms to beat human world champion G. Kasparov in the chess game. To deal with uncertainty, *admissible heuristics* can be used.

Learning: Reinforcement learning solves the problems by looking into every state, accommodating uncertainty, without the system designer examining all scenarios. If a robot directly executes a sequential decision making algorithm, it is called *online learning*. If a simulator of the environment is available for training examples, it is known as *offline learning*.

▶ **Exercise (Pole Balancing):** Figure 5.3 depicts a classic example of reinforcement learning for a robot to balance the pole. Please define the state of

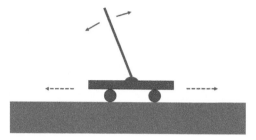

Figure 5.3 Pole balancing.

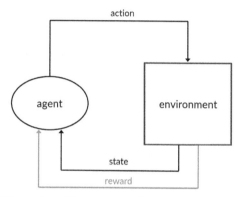

Figure 5.4 Illustration of reinforcement learning.

this reinforcement learning. Can you assign reward to complete the modeling of pole balancing as a reinforcement learning problem.

▶ Exercise (8-Queen): Please use RL to solve the 8-queen problem in Chapter 2.

5.1.3 Reinforcement Learning Based on Markov Decision Process

One extensively used formulation for studying RL is Markov decision processes (MDP). Figure 5.4 shows the basic elements of RL and the interactions between them. Robot, often referred to as agent, takes actions and observes rewards from the environment. Such consecutive decisions made by robot is well captured by the framework of MDP. In MDPs, the interaction between the decision maker (i.e. agent or robot) and environment is described by *states*, *actions*, and *rewards*. Agent observes the information pattern of the environment and establish an awareness of its own state. According to its

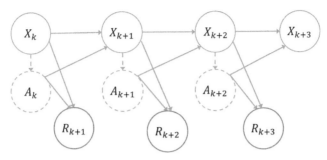

Figure 5.5   MDP formulation of reinforcement learning.

current state, it possesses some choices of actions that will transits itself to another state and then receive some real-valued reward from the environment. A sequence of state-action-reward happens in a so-called *decision epoch*. An MDP consists of a number of decision epochs, which we call the MDP's *horizon length*. The goal of the agent, and therefore the objective of MDP analysis, is to determine a rule for each decision epoch for selecting an action such that the collective rewards are optimized.

To form MDP in a mathematical way, let $\{S_t \in \mathcal{S}\}$ be the sequence of states at epoch $t \in \{0, 1, 2, \ldots, K\}$ where $K \leq \infty$ is the horizon length and $\mathcal{S}$ is the state space. Action taken at decision epoch t is denoted by a_t which belongs to the action space $\mathcal{A}$. More precisely, the action space can depend on the current state, denoted by $\mathcal{A}(s_t)$. The real-valued reward R_{t+1} is accrued after epoch t. Since the reward is given on the basis of last state S_t and action A_t, the notation for reward R_{t+1} (instead of R_t) is used because it captures the fact that after certain epoch t, reward R_{t+1} and the consequential state S_{t+1} are determined together. Figure 5.5 illustrates the relationships between states, actions, and rewards. The direction of arrow indicates on which part it would make impacts. The figure tells that, as we have described, states and actions influence the returning reward and the next state jointly.

For a finite MDP, of which the number of states and actions are all finite, its dynamics at any time t can be characterized by these arguments through a discrete probability function $p : \mathcal{S} \times \mathcal{A} \times \mathcal{S} \times \mathcal{R} \rightarrow [0, 1]$. To be specific, for any particular value $s, s' \in \mathcal{S}, r \in \mathcal{R}, a \in \mathcal{A}(s)$,

$$p(s', r|s, a) \doteq Pr\{S_{t+1} = x', R_{t+1}|S_t = x, A_t = a\}.$$

From it, two frequently used functions can be computed. One is the *state-transition probabilities*,

$$p_{ss'}^a \doteq Pr\{S_{t+1} = s'|S_t = s, A_t = a\} = \sum_{r \in \mathcal{R}} p(s', r|s, a).$$

We often call s' the successor state and another terminology is the *expected rewards* given any state-action pair,

$$r(s,a) \doteq \sum_{r \in \mathcal{R}} r \sum_{s' \in \mathcal{S}} p(s', r|s, a).$$

In the process of deciding actions, the agent follows specific decision rules $\{\delta_t\}$ corresponding to each epoch $t \in \{0, 1, \ldots, K\}$ that tells which action it should take with respect to its state. That is, the decision rule δ_t is a function mapping every $s \in \mathcal{S}$ onto an action $a \in \mathcal{A}$. In general, we refer to the sequence of decision rules as *policy* and denote it by $\pi = \{\delta_0, \ldots, \delta_K\}$. If the decision rules are the same in spite of epochs, the policy is *stationary* and we denote it by $\bar{\pi} = \{\delta, \delta, \ldots\}$.

The goal of MDP analysis is to derive an optimal policy π^*, where optimal means that no matter at which state the agent is, executing the policy will lead to the maximal expected future rewards.

$$\pi^* \doteq \arg\max_{\pi} v_{\pi}(s) \quad \forall s \in \mathcal{S} \tag{5.9}$$

where $v_{\pi}(s)$ is defined to be the *state-value function* (or *value function* for short) for the agent who follows policy π at state s. The criterion to measure the value is simply the expected reward in future, i.e.

$$v_{\pi}(s) \doteq \mathbb{E}_{\pi} \left[\sum_{i=t+1}^{K} R_i \middle| S_t = s \right]$$

In many cases where it is not clear how long the MDP will goes, considering infinite horizon length is more appropriate. To ensure the value function converge, a discount factor $0 \leq \gamma < 1$ is required. Hence, a general form of the value function can be written as

$$v_{\pi}(s) \doteq \mathbb{E}_{\pi} \left[\sum_{i=t+1}^{K} \gamma^{i-t-1} R_i \middle| S_t = s \right] \tag{5.10}$$

where discount factor is subject to $0 < \gamma < 1$. Despite state-value function, we can also evaluate state-action pairs (s, a) for policy π. Denoted by $q_{\pi}(s, a)$, it is referred to as *action-value function for policy* π,

$$q_{\pi}(s, a) \doteq \mathbb{E}_{\pi} \left[\sum_{i=t+1}^{K} \beta^{i-t-1} R_i \middle| S_t = s, A_t = a \right]. \tag{5.11}$$

If we use another notation G_t to denote the sum of (discounted) rewards received after epoch t

$$G_t = \sum_{i=t+1}^{K} \gamma^{i-t-1} R_i,$$

the value function becomes

$$v_\pi(s) \doteq \mathbb{E}_\pi \left[\sum_{i=t+1}^{K} \gamma^{i-t-1} R_i \middle| S_t = s \right] \tag{5.12}$$

$$= \mathbb{E}_\pi \left[R_{t+1} + \gamma \sum_{i=t+2}^{K} \gamma^{i-t-2} R_i \middle| S_t = s \right] \tag{5.13}$$

$$= \mathbb{E}_\pi \left[R_{t+1} + \gamma G_{t+1} \middle| S_t = s \right] \tag{5.14}$$

The second term in the last row is in fact related to the action taken at that epoch and the next state. If we let $\pi(a|s)$ be the probability of choosing action $A_t = a$ at state $S_t = s$; let the following reward $R_{t+1} = r$ and successor state $S_{t+1} = s'$ determined by function $p(r, s'|s, a)$, equation (5.14) turns into

$$\sum_{a \in \mathcal{A}(s)} \pi(a|s) \mathbb{E}_\pi \left[R_{t+1} + \gamma G_{t+1} \middle| S_t = s, A_t = a \right]$$

$$= \sum_{a \in \mathcal{A}(s)} \pi(a|s) \sum_{s'} \sum_{r} p(r, s'|s, a) \left[r + \gamma \mathbb{E}_\pi [G_{t+1}|S_{t+1} = s'] \right]$$

$$\Rightarrow v_\pi(s) = \sum_{a \in \mathcal{A}(s)} \pi(a|s) \sum_{s'} \sum_{r} p(r, s'|s, a) \left[r + \gamma v_\pi(s') \right] \tag{5.15}$$

Equation (5.15) is the *Bellman equation for* $v_\pi(s)$, which implies the relationship between a state's value function and its successor state's value. Something nice to have this form of value function is that, at every decision epoch t for any state s_t, the optimal decision rule δ_t is to choose the action that will maximize the expected reward plus the discounted value of the next state.

5.1.4 Bellman Optimality Principle

Solving a reinforcement learning task means, roughly, finding a policy that achieves a lot of reward over the long run. Therefore, a policy π is said to be better or equal to a policy π' if its expected return is greater than or equal to that of policy π' for all states. That is, $\pi \geq \pi'$ if and only if $v_\pi(s) \geq v_{\pi'}(s)$ for all $s \in S$. There might be more than one optimal policy but we denote all of them by π_*. And they share the same state-value function, which is called *optimal state-value function*, denoted by v_*, and is defined as

$$v_*(s) = \max_\pi v_\pi(s), \quad \forall s \in S.$$

Optimal policies also share the same *optimal action-value function*

$$q_*(s, a) = \max_\pi q_\pi(s, a), \quad \forall s \in S \text{ and } a \in A(s).$$

Substitute the optimal value function into Bellman equation (5.15), we get the *Bellman optimality equation* for a state s,

$$v_*(s) = \max_{a \in A(s)} q_{\pi_*}(s, a)$$

$$= \max_a \mathbb{E}_{\pi_*}\left[R_{t+1} + \gamma G_{t+1} \Big| S_t = s, A_t = a\right]$$

$$= \max_a \mathbb{E}[R_{t+1} + \gamma v_*(S_{t+1}) \Big| S_t = s, A_t = a]$$

$$= \max_a \sum_{s',r} p(s', r|s, a)[r + \gamma v_*(s')]$$

Intuitively, Bellman optimality equation expresses the fact that the value of a state under an optimal policy must equal the expected return for the best action from that state.

Once we obtain the optimal state-value function v_*, it is easy to derive the optimal policy. There are three propositions that would hopefully lead to the solution of v_*. The first states that in finite MDP there exists only one optimal value function, that is, there is a unique fixed point within the set of real-valued functions. The second and the third propositions provide basis for two general methods to obtain the optimal policy: *value iteration* and *policy iteratioin*.

Proposition 2 (Unique Solution of Bellman Equation). *Given the reward and state transition probability $p(r, s'|s, a)$, the optimal value function v^* is the unique solution of Bellman equation.*

Proposition 3 (Convergence of Value Iteration). *Any sequence $\{v_t\}$ starting from any $v_0 \in V$ and updated by*

$$v_{t+1}(s) = \max_{a \in A(s)} \sum_{s'} \sum_r p(r, s'|s, a)\left[r + \beta v_t(s')\right], \quad \forall x \in S \quad (5.16)$$

will converge to v^*.

Proposition 4 (Convergence of Policy Iteration). *Starting from any station-ary policy $\bar{\pi}_0$, if the value functions are updated through*

$$v_{\bar{\pi}_i}(s) = \sum_{a \in \mathcal{A}(s)} \bar{\pi}_i(a|s) \sum_{s'} \sum_r p(r, s'|s, a)\left[r + \beta v_{\bar{\pi}_i}(s')\right], \quad \forall i = 1, 2, \dots,$$

(5.17)

after the value of all states converge, improve upon the policy to construct another stationary one $\bar{\pi}_{i+1} = \{\delta_{i+1}, \delta_{i+1}, \delta_{i+1} \dots\}$ by

$$\delta_{i+1}(s) = \arg \max_{a \in \mathcal{A}(s)} \sum_{s'} \sum_r p(r, s'|s, a)\left[r + \beta v_{\bar{\pi}_i}(s')\right].$$

(5.18)

The sequence of value function $\{v_{\bar{\pi}_i}\}$ will converge to the unique optimal one v^*.

▶**Exercise (Pole Balancing):** In Figure 5.3, suppose we only consider the scenario as a plane, which means that the platform can only move right and left with possible speed at $0, 1, 2, 3, 4, 5$ m/sec, and the pole can only move clockwise or counter-clockwise. Assume the platform can precisely know the angle of the pole that has uniform density (and thus weight distribution). Please design a reinforcement learning algorithm to balance the pole.

5.2 Q-Learning

Reinforcement learning can proceed in two categories: *model-based learning* and *model-free learning*. In model-based learning, the agent tries to estimate the parameters about the environment to fit a model of the environment's dynamics. Models such as neural network, Gaussian process are often applied toward implementation. With the approximated function of system, dynamic programming methods such as value iteration and policy iteration can be applied to compute the optimal policy.

5.2.1 Partially Observable States

In many cases of engineering interest, the agent does not exactly know the state, while relies on certain mechanism (e.g. sensors to observe) to estimate the state. The following figure illustrates the scenario, where b is the estimate of the state. For example, a robot to clean a room. Such a new setting is very much like a MDP, except that after taking an action a_k, the new state x_{k+1} is unknown, but we have an observation y_{k+1} that can be treated as a stochastic

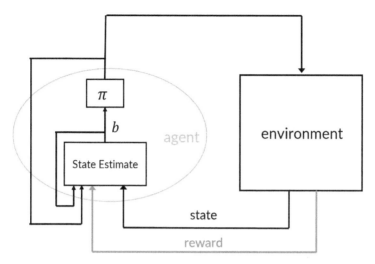

Figure 5.6   Partially observable reinforcement learning.

function of x_k and a_k: $p(y_{k+1}|x_k, a_k)$. This is called *partially observed MDP* (POMDP). When $y_{k+1} = x_{k+1}$, POMDP reduces to MDP. Please note that we can infer the state in a randomized manner.

However, the Markov property does not necessarily hold for observations, even the state transition follows a Markov process. At any time, the agent may calculate the most likely state and take an action accordingly (or, an action to gather information further). To maintain the Markov process, the agent keeps an internal belief state b_k that summarizes its experience. The agent uses a state estimator to update the belief state b_{k+1} based on last action a_k, current action a_{k+1}, previous belief b_k. A policy π generates next action a_{k+1} based on current belief state (please compare to the actual state in model-based approach). The belief state has a probability distribution over the states of the environment given the initial belief state (prior to any action) and the past observation-action history of the agent. The spirit of learning can thus involve such belief state-action pairs, instead of the actual state-action pairs, by

$$Q(b_t, a_t) = \mathbb{E}\left[r_{t+1}\right] + \gamma \sum_{b_{t+1}} P(b_{t+1} \mid b_t, a_t)V(b_{t+1}) \qquad (5.19)$$

Partially observed formulation of reinforcement learning coincides with realistic design for autonomous agents and machines such as robots or autonomous vehicles.

5.2.2 Q-**Learning Algorithm**

Q-learning was devised by Watkins that provides a learning method for agent when facing a task without initially knowing the dynamics of environment, i.e. state transition probability and reward function in 1992. Watkins classified Q-learning as incremental dynamic programming, because of the step-by-step manner in which agent determines the optimal policy.

In Q-learning, agent updates the action-value function every decision epoch depending on the maximum action-value in the successor state. It only uses the immediate payoff and based on the best it can do in the next state to refine its action-value function $q(s, a)$. The learning proceeds in the following steps:

(i) Arbitrarily initialize $q(s, a)$ for all $x \in \mathcal{S}$ and $a \in \mathcal{A}$. (If s is terminal state, its value must be zero.)
(ii) Choose an action a in current state s based on current Q-value (i.e. action-value) estimates, e.g. ϵ-soft policies.
(iii) Take action a, then observe reward r and next state s'.
(iv) Update Q-value following

$$Q(s, a) \leftarrow Q(s, a) + \alpha[r + \gamma \max_{a'} Q(s', a') - Q(s, a)]$$

Note that $0 \leq \alpha < 1$ is the learning rate. A higher value means that the refinement of Q-value occurs quickly and hence the learning goes faster. γ denotes discount factor, restricted by $0 < \gamma < 1$, implies that future rewards are worth less than immediate rewards.

Example: As shown in Figure 5.8, black squares in the figure are the obstacles. A rat can only pass through white squares to reach the food (the rightest

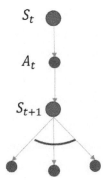

Figure 5.7   Backup diagram for Q-learning.

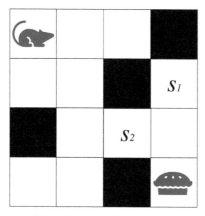

Figure 5.8 Maze problem: How can the rat find the way out to the food?.

square of the bottom row), how do the rat find the way to the food under a restricted environment? This is a simple maze problem. Given an entry and an exit point, how to find the path between the entry and exit? If there exists multiple ways, how to find the shortest path?

Solution: The maze problem can be solved using reinforcement learning. In this model, the rat has exactly 4 different actions: {*up, down, left, right*}, which is the action space. The state space consists of all the cases the rat located in each white square. In total we have 12 different states (we only have 12 white squares).

To encourage the rat to find the shortest path to the food, we need to assign a rewarding scheme: (i) The rewards after each action will be a floating point ranging from -1.0 to 1.0 in our model. When the reward for an action is positive, the action is "encouraged" and "decouraged" ortherwise; (ii) An action that leads to a blocked square will cost a negative reward of 0.5 such that hopefully the rat can find a path without passing through any blocked square; (iii) To learn the shortest path, a negative cost of 0.05 is given for moving from one square to another square to avoid wandering around; (iv) If the rat is in the margin square of the figure, actions that will lead to outside of the maze boundary will cost a negative reward of 0.5; (v) Actions that will result in new states that already visited will be penalized by 0.25; (vi) The 'food' state is the ultimate goal the rat wants to reach, which will have a positive reward of 1.

In the exploration/learning process, the rat will try different actions to learn the rewards. We give four example exploration episodes in Figure 5.9.

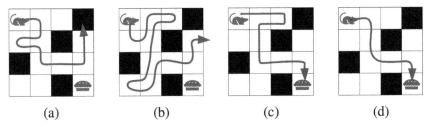

| (a) | (b) | (c) | (d) |

Figure 5.9   Four different example exploration episodes of maze problem.

We can get the rewards of the four exploration episodes as:

(a): $r_a = -0.05 \times 9 + (-0.25) + (-0.5) = -1.2,$

(b): $r_b = -0.05 \times 15 + (-0.25) \times 4 + (-0.5) = -2.25,$

(c): $r_c = -0.05 \times 8 + (-0.25) + 1 = 0.35,$

(d): $r_d = -0.05 \times 6 + 1 = 0.7.$

After each episode, we can update the average reward for each visited state. The reward for a state is defined as the total reward of the following exploration. For example,

- When the rat is located in state s_1 as shown in Figure 5.8, the initial reward is set as 0, after episode (a), the reward $R_{s_1} = -0.5$. After episode (b), $R_{s_1} = -0.5 + (-0.5) = -1$. Episode (c) and (d) did not visit s_1, thus the final reward after four episodes will be -1, the state is "decouraged" to visit in the following episodes.
- When the rat is located in state s_2, the reward $R_{s_2} = -0.05 \times 3 + (-0.5) = -0.65$ after episode (a), $R_{s_2} = -0.65 + (-0.05) \times 3 + (-0.5) = -1.3$ after episode (b), $R_{s_2} = -1.3 + (-0.05) \times 2 + 1 = -0.4$ after episode (c), $R_{s_2} = -0.4 + (-0.05) \times 2 + 1 = 0.5$ after episode (d), thus state s_2 is "encoraged" after four episodes in the following exploration.

After a large number of trials, the rat can learn an estimated reward value on each feasible state, the shortest path can be found by moving to the next state that has the maximum reward value, here in our example, is the path shown in episode (d).

5.2.3 Illustration of Q-Learning

Q-learning is widely applied in many engineering scenario. In this sub-section, we apply Q-learning to model an scenario of intelligent transportation as a multi-agent system (MAS) while each agent behaves in reinforcement learning.

Under the ultimate safety requirement, given destination and street map, an AV operates in a way to identify the best route under certain criterion, such as shortest, fastest, energy efficient, etc. However, an AV must operate with other AVs in different fleets and other human driving vehicles, without priori knowledge about the presence of other vehicles on the streets and their destinations. In AI, an AV can be viewed as an agent of RL. We therefore are dealing with the multi-agent system sharing the common resources (i.e. streets).

Figure 5.10 illustrates the multi-agent system of AVs over the Manhattan Street Model of square grid topology. In the Figure 5.10, the topology is X blocks ($X = 4$) in lengthwise, Y blocks ($Y = 6$) in crosswise, while the length of each block is b ($b = 5$). An AV is viewed as an agent of RL to the optimal policy toward the destination, while the agent knows the street map and its destination but does not know other vehicles on the streets until within immediate sight.

The navigation of each AV can be represented by the rules as follows:

- The road has 2 lanes (one lane for each direction)
- Select an action a_k from the action space at the state s_k at time k
- At the intersection, each vehicle should stop for one step (at the state with red circles from the right part of Figure 5.11) to check whether other vehicles coming from different directions (i.e. as the function of stop sign)

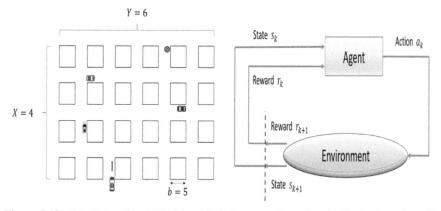

Figure 5.10   Manhattan Street Model and Reinforcement Learning for Navigation of an AV.

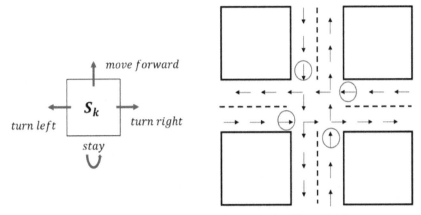

Figure 5.11   Action Pattern (left) and Action Flow (right).

- During moving forward, if there is another vehicle in front, both vehicles keep one step separation

Reward

We define the vehicle's position at time k as state s_k; $s_k = \{l_x, l_y\} \in S$, and $l_x = 1, 2, \ldots, L_x = Xb + 2(X - 1)$, $l_y = 1, 2, \ldots, L_y = Yb + 2(Y - 1)$, a_k represents the possible action at state s_k at time k, $a_k \in \{forward, left, right, stay\} = \mathcal{A}$. Suppose vehicles are indexed, $i = 1, 2, \ldots, N$, where N is total number of vehicles coming into the Manhattan street region. The left part of Figure 5.11 shows the picture of the action pattern of the vehicle in the street. The right part of Figure 5.11 represents the example of the possible actions for each state (position). The agent decides the action based on the RL with *reward* r that is a value used for the agent's decision. The reward map $R_{i,k}$ for i-th vehicle at time k is the matrix of reward $r_{s_k} = r_{l_x, l_y}$ constructed on the Manhattan street map centered in the location (l_x, l_y).

$$R_{i,k} = [r_{s_k}] = \begin{bmatrix} r_{0,0} & \cdots & & \cdots & \cdots & r_{0,L_y} \\ \vdots & \ddots & & \cdots & & r_{1,L_y} \\ \vdots & & \vdots & r_{l_x,l_y} & & \vdots \\ \vdots & & \vdots & & \ddots & \vdots \\ r_{L_x,0} & r_{L_x,1} & & \cdots & \cdots & r_{L_x,L_y} \end{bmatrix}$$

The reward (or penalty) at state $s_k = \{l_x, l_y\}$ takes one of the following values to represent different levels of desirability to complete the mission (i.e. safely reaching destination).

$$r_{s_k} = r_{l_x, l_y} = \begin{cases} R_{destination} & arriving\ destination \\ R_{prohibit} & motion\ being\ prohibit \\ R_{another} & another\ agent\ passing \\ R_{step} & agent\ passing\ s_n \\ 0 & otherwise \end{cases} \quad (5.20)$$

where $R_{destination}$ is usually a significant positive value for the successful completion of agent's mission; $R_{prohibit}$ is set as $-\infty$; $R_{another}$ is set to be a negative value due to disfavored to complete agent mission and to loss of efficiency of the entire public infrastructure (i.e. losing efficiency of streets); R_{step} is typically a small negative value for time and energy consuming. Such setup allows further development of RL later.

Unlike the supervised machine learning for data analytics that the data model is trained based on labeled datasets, RL explores by interacting with the environment, which is more suitable for learning process of smart agents like robots and autonomous driving vehicles. The concept is that the agent gets rewards based on actions interacting with the environment and then makes a decision such as the environment will give large rewards. In the Manhattan Street Model problem, we define the RL parameters as below.

- Agent: The machine of AI to learn and to take action a_k at time k
- Environment: The situations of the roadways which the vehicle observes, position s_k at time k in this illustration
- Reward r_k: A real number to value the action a_k at time k, based on the position s_k
- Value $V(s_k)$: The total reward that the agent receives over the long run from s_k
- Policy π: The series of actions by the agent from time k to the horizon, typically with discount for future actions

Value Function

The goal of RL is to find optimal policy which maximizes the value function. The value function indicates that how good it is for the agent in a given state, and the value function of a state $s_k = s$ under the policy π, denoted $v_\pi(s)$, is the expected return when starting in and following π. G_k is defined as some specific function of the reward sequence and with future time step $k + d$, G_k

can be looked the basic potential value of state s_k for $k+d; d=1,2,\ldots.$ The future rewards will be discounted by a parameter $\gamma, 0 \leq \gamma \leq 1$, representing the discount rate.

$$G_k = \sum_{d=0}^{\infty} \gamma^d r_{k+d+1} \tag{5.21}$$

$$v_\pi(s) = E_\pi[G_k|s_k = s] = E_\pi[\sum_{d=0}^{\infty} \gamma^d r_{k+d+1}|S_k = s] \tag{5.22}$$

The state-action value function indicates the expected reward starting from current to the horizon of depth D, taking the action, and thereafter following policy $\pi = \{a_k, \cdots, a_{k+D}\}$.

$$q_\pi(s,a) = E_\pi[G_k|s_k = s, a_k = a]$$
$$= E_\pi[\sum_{d=0}^{\infty} \gamma^d r_{k+d+1}|s_k = s, a_k = a] \tag{5.23}$$

The optimal value-function $v_*(s)$ and state-action value $q_*(s,a)$ can be obtained as below [3].

$$v_*(s) = \max_\pi q_{\pi*}(s,a) \tag{5.24}$$

$$q_*(s,a) = \max_\pi q_\pi(s.a) = E[r_{k+1} + \gamma v_*(s_{k+1})|s, a_k = a] \tag{5.25}$$

Q-Learning

Since the agent is unlike to know the exact state, it must observe to generate the belief of the state for actions. By replacing the state-action to the belief-action, we obtain a popular variant of RL, *Q-Learning*. The learned action-value function Q directly approximates the optimal action-value function q_*.

$$Q_{k+1}(s,a) \leftarrow Q(s,a) + \alpha[r_{k+1} + \gamma \max_a Q_{k+1}(s',a) - Q_k(s,a)] \tag{5.26}$$

where a constant step-size parameter α scales how the newly obtained information replaces the old information. The Q-learning algorithm is summarized as the Algorithm 1 [3].

Navigation

While an AV heading to the destination, the jamming (two or more vehicles simultaneously heading to the same position) occurs more likely as the number of AVs is increasing, especially competing resource at the intersection

---

Algorithm 1: Q-value for each state

---

1   <u>function</u> $Q_k(s,a), s \in \mathcal{S}, a \in \mathcal{A}$;

2   Initialization $k = K, Q(s,a) = 0$ **while** $k \geq 0$ **do**

3 | Observe r, s'(Possible previous state) for taking action a;

4 | $Q_k(s,a) \leftarrow Q_k(s,a) + \alpha[r + \gamma \max_a Q_{k+1}(s',a) - Q(s,a)]$;

5 | $k \leftarrow k - 1$;

6 **end**

---

without knowing the future actions of each other. If we do not resolve the jamming, it can result in a collision such as an AV hitting another vehicle/bicycle. We therefore have to modify the reward distribution of RL for an AV navigating on the Manhattan Streets. The learning procedure depends on whether there exists wireless communication or not. When the agent applies the learning with communication, it is in the communication mode, otherwise, without communication mode. Assuming there are several AVs navigating on Manhattan Streets, and we will look into MAS in different situations; i) no communication, ii) Ideal V2V Communications that allows information exchange among AVs/agents, and iii) Ideal V2I2V Communications to alleviate scaling concern, which any communication must be in two hops, V2I and I2V. Given safety criterion (no collision nor crashing), the expected duration to navigate through the Manhattan streets for each AV serves the performance index of such MAS. Based on these benchmark investigations, further practical issues of wireless communication will be taken into account in Chapter 10.

If there are no other vehicles on the streets, the single AV directly heads to the destination without any interference. In the Manhattan Model street, given the street map and destination, each agent applies Q-learning to select the optimal route toward the destination for next episodes within the depth of horizon. Q-leaning proceeds navigation by making actions from the neighbor states of next possible state s' where the rewards from every state can be received.

When there are multiple AVs on the streets, each AV has no information about other vehicles. Therefore, each vehicle has to stop before entering the intersection, and to observe other vehicles from different directions. At this time, the agent observes the positions of other vehicles, but cannot get the information about where to head. Therefore the agent shall predict and calculate the expected rewards of them. Each AV recognizes other vehicles by the following observations.

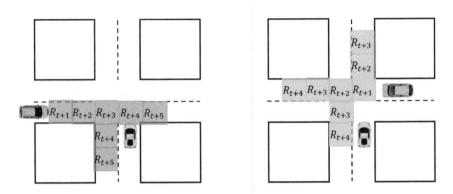

Figure 5.12   Predicting future rewards.

- While i-th vehicle driving in the straight street, the vehicle can see the vehicle in front
- Before entering the intersection, each vehicle should stop for one step and observes other vehicles ($j \in \mathbb{I}_k$) from different directions

When the agent (i.e. AV) observes the j-th agent at s_k, the agent sets the negative reward $r_{s_k} = R_{stay}$ because other agents are regarded as the obstacles for driving. However, even the agent recognizes other agents' positions, the agent still does not know how they move and has to wait until resolution as the stop sign in modern traffic regulations.

When the vehicle is at $s_k = s$, the possible action $a_k = a$, and the next state is $s_{k+1} = s'$, there are several patterns of $a_k = a, s_k = s$. For s_{k+d}, the expected reward $r_{s'}$ is therefore

$$
\begin{aligned}
r_{s'} &= \mathbb{E}[r_{s'}|s_{k+d-1} = s'] + \mathbb{E}[r_{s'}|s_{k+d-1} = s] \\
&= p_{stay} r_{s'} + \sum_a p_a r_s
\end{aligned}
\tag{5.27}
$$

where p_{stay} is the probability of the vehicle staying at the same position in next time instance, p_a is the probability of the vehicle taking possible action $a \in \mathcal{A} - \{stay\} = \{forward, left, right\}$.

Therefore, we use Q-learning to model an agent of a MAS to represent the behavior of autonomous vehicles on Manhattan Streets. This illustration will be developed further in later chapter about MAS.

■ **Exercise:** A robot is placed to walk through a maze as Figure 5.13 by starting from the exit in blue and getting out from the exit in red. The robot walks

Figure 5.13 Maze walking.

from one square to an immediate neighboring square (i.e. moving up, down, right, left). The black squares mean blocking (i.e. prohibited for a robot to get in, say a wall). The white squares mean the robot being allowed to move. Each time duration, a robot can sense its alternative movements (maybe more than one) and select its permissible movement in a random manner.

(a) Please develop and algorithm to walk the maze. *Hint: This algorithm can be developed by dynamic programming, MDP, or RL.*

(b) Then, repeat the reverse direction (enter from red exit and leave the blue exit). What is the average steps for these two walks on this maze? *Hint: This is for fairness in performance evaluation.*

(c) Suppose the sensing (black square or not) of a robot has probability of error e. Up to now, we assume $e = 0$. If $0 < e \ll 1$, while the robot mistaking a black square to be white will waste one unit of time duration. Please repeat (b) for $e = 0.1$. *Hint: The modifications of the algorithm in (a) would be different for dynamic programming, MDP, RL.*

5.3 Model-Free Learning

Reinforcement learning can proceed in two categories: *model-based learning* and *model-free learning*. In model-based learning, the agent tries to estimate the parameters about the environment to fit a model of the environment's dynamics. Models such as neural network, Gaussian process are often applied toward implementation. With the approximated function of system, dynamic programming methods such as value iteration and policy iteration can be applied to compute the optimal policy. However, in many cases, the models may not be available, or at least with uncertainty. It motivates the development of model-free approach, which machine learning is generally favored nowadays. For robotics, it is often true that the precise models are not available.

Instead of analyzing how the environment will respond to the agent's action like model-based methods, model-free learning methods directly learn value function (or action-value function) from experience. A straightforward way to estimate the value of one state or the value of an action is to average the overall historical rewards for that state or that action, which is called *Monte Carlo method*. It facilitates another significant and novel RL known as *temporal-difference (TD) learning*.

5.3.1 Monte Carlo Methods

> **Box (Monte Carlo Simulation):**
> If a function $h(\mathbf{x}) \in \mathbb{R}^n$ is hard to compute, *Monte Carlo simulation* can be applied to generate independent random samples on $\mathbb{R}^n$, and then to take average for the outcomes. By the law of large number, we can obtain a good approximation. For example, to estimate the value of π, N_{mc} samples are randomly selected from $[0, 1]^2$, and count the number of samples giving $x_1^2 + x_2^2 \leq 1$ as N_π. $\frac{N_\pi}{N_{mc}}$ shall approach the true value of π.

For better estimation of state-value or action-value, Monte Carlo methods divide problems into episodic tasks. Only when a terminal state is reached, the rewards along the trajectory of states will be updated. As the number of state x being visited goes to infinity, $v_\pi(x)$ will converge to the optimal value function $v^*(x)$. This can be verified easily since each return is an independent, identically distributed estimate of $v_\pi(x)$ with finite variance, by law of large numbers the sequence of averages of these estimation will converge to its expected value. However, to ensure that every state-action pair is visited infinite times, the agent must maintain exploration so that not only specific actions will be chosen and estimated their values. Two ways to continue exploration are:

- exploring starts: Every state-action pair has a non-zero probability of being the starting pair.
- soft policies: In a soft policy π, every action has a non-zero probability to be chosen, that is

$$\pi(a|x) > 0, \quad \forall x \in \mathcal{S}, \forall a \in \mathcal{A}.$$

As one could imagine, exploring starts are sometimes not useful, particularly when an agent is learning from actual interaction with the environment. Hence, the most common approach to promise all state-action pair will be encountered infinite times is soft policies. There are two attempts to imply soft policies: *on-policy MC methods* and *off-policy MC methods*. On-policy means that the soft policy that is used to maintain exploration will also be the repeatedly improved policy (ultimately becomes the optimal policy). Off-policy separates the soft policy, according to which explorable actions are selected, and the policy being improved. In the terms of off-policy methods, the soft policy is usually referred to as *behaviour policy* while the continuously refined one is called *target policy*.

On-policy first-visit Monte Carlo Methods:

(i) Arbitrarily initialize $Q(x, a)$ for all $x \in \mathcal{S}$ and $a \in \mathcal{A}$. (If x is terminal state, its value must be zero.)
(ii) Let $Returns(x, a)$ be an empty list. Let π be an arbitrary soft policy.
(iii) Generate an episode using π.
(iv) For each state-action pair (x, a) appears in the episode, append the return G that follows the first appearance on $Returns(x, a)$.
(v) Update $Q(x, a)$ by using the average of $Returns(x, a)$ for all $x \in \mathcal{S}$ and $a \in \mathcal{A}$.
(vi) For each state $x \in \mathcal{S}$, define $a^* = \arg\max_{a \in \mathcal{A}(x)} Q(x, a)$. (If more than one a^* exist, randomly select one of them.) Update the decision rule by

$$
\pi(a|x) = \begin{cases} \frac{\epsilon}{|\mathcal{A}(x)| - 1} & \text{if } a \neq a^* \\ 1 - \epsilon & \text{if } a = a^* \end{cases}
$$

(vii) Goes back to (ii). (This algorithm should run significantly large times in order to meet the assumption of infinite visits.)

In off-policy methods, *importance sampling* is usually utilized to form a connection between the behaviour policy π_b and the target policy π. This is a general technique used when we are estimating properties of one distribution under samples from other distributions. Assume an episode starts from epoch t to T, the importance-sampling ratio which stands for the relative probability of the trajectory under π_b and π is

$$
\rho_{t:T} = \frac{p(X_t, A_t, \ldots, X_T, A_T | X_t, A_{t:T} \sim \pi)}{p(X_t, A_t, \ldots, X_T, A_T | X_t, A_{t:T} \sim \pi_b)}
$$

$$= \frac{\prod_{i=t}^{T} \pi(A_i|X_i)p(X_{i+1}|X_i, A_i)}{\prod_{i=1}^{T} \pi_b(A_i|X_i)p(X_{i+1}|X_i, A_i)}$$

$$= \frac{\prod_{i=t}^{T} \pi(A_i|X_i)}{\prod_{i=1}^{T} \pi_b(A_i|X_i)}. \tag{5.28}$$

It turns out that the importance sampling does not count on the MDP, instead the two policies and the sequence of states and actions effect. Then, the value function $v_\pi(s)$ can be estimated by *ordinary importance sampling*

$$v(s) \doteq \frac{\sum_{t \in \mathcal{T}(x)} \rho_{t:T} G_t}{|\mathcal{T}|} \tag{5.29}$$

or *weighted importance sampling*

$$v(s) \doteq \frac{\sum_{t \in \mathcal{T}(x)} \rho_{t:T} G_t}{\sum_{t \in \mathcal{T}(x)} \rho_{t:T}} \tag{5.30}$$

$\mathcal{T}(x)$ is the set of time step that state x is visited in this episode. (But we only consider first visit MC so actually there is only one element in the set.) G_t denotes the sum of return from time t up to T.

Example: (Tic-Tac-Toe) Tic-tac-toe is a game in which two players mark on the 3×3 board in turns. Who makes three in a row, column, or diagonal wins the game. Since how well a player plays is determined at the end of the game while all the previous moves have their impact, this can be modelled into an episodic task. Moreover, the returns are only given when terminal states happen–win, lose, or tie. To apply reinforcement learning, first define the state x_k as the game board at the k-th turn of play. The action set for x_k is all the empty grids at that moment. Second, the reward setting in RL is critical since rewards serve as a means for computers to complete the goal. In tic-tac-toe game, we wish the computers to be good players that strive to win or at least not lose. Hence we can set up the rewards for player

$$r = \begin{cases} 1 & \text{if he wins} \\ -100 & \text{if he loses} \\ 0 & \text{if game ties.} \end{cases}$$

For simplicity, we label every possible state, e.g. s_0 for the empty grid, s_1 for the board on which top-left grid is marked by cross etc. (As shown as figure below.)

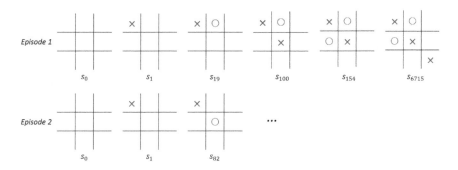

At every iteration, we simulate a tic-tac-toe game, i.e. an episode, generating a sequence of board state such as $s_0, s_1, s_{19}, s_{100}, \ldots$. The sequence of state ends in a terminal state by then a reward is given. The MC algorithm will update the action value function for those who had happened in this sequence, as a sample of the true value function. In episode 1, if we say the action of marking the top-left grid is a_1, then its action value function is updated by $q(s_0, a_1) = 1$ due to his win.

Running the on-policy first-visit MC algorithm for significantly large times of iteration, all actions and states are promised to be operated and occur for considerable times. Even without opponent's strategy in mind (although we do have a random mechanism for generating the opponent's movements when simulating), MC algorithm can eventually make good predicts on action value. Thereby, the computer can based on the action value function to make an optimal move.

5.3.2 Temporal Difference Learning

Q-learning uses an immediate reward to update the action-value function. However, sometimes a longer update horizon is essential when reward can only be obtained after some time delay or after transiting into certain states, which are called *terminal states*. The sequence of states that ends with terminal states is called *episodes*, whose length is defined as T. For example, if the task is to find a path through the maze to the other end, the episode might end in two ways: reaching the exit or going to a dead end.

Methods that use the rewards from the start to the end of episode to update state values are is called *Monte Carlo methods*. It can be considered as learning from completely raw experience. And the update length is regarded as infinity since it only refines state value when an episode ends. Nonetheless, for some tasks it is not necessary to wait until the end of all episodes. Instead, an agent can evaluate its action after n decision epochs. This approach is called *n-step temporal difference* (TD-n) methods. TD-n learning is widely adopted to deal with the cases of delayed rewards (i.e. rewards can not be obtained immediately).

Remark: TD learning can be viewed as a combination of Monte Carlo and dynamic programming (DP) ideas. Similar to Monte Carlo methods, TD methods learn directly from the raw experience without a model of the environment dynamics. Also like DP, TD learning updates estimates without waiting for a final outcome/reward, as it bootstraps (i.e. estimates based on estimation) like a predictor.

Remark (TD Prediction): Given some experience following a policy π, both update the estimate value of v_π for the non-terminal states S_t occurring in that experience. Monte Carlo waits until the return following the known visit, then uses that return as a target for $V(S_t)$. A simple every-visit Monte Carlo for non-stationary environments is

$$V(S_t) \leftarrow V(S_t) + \alpha \left[G_t - V(S_t) \right] \tag{5.31}$$

where G_t is the actual return, α is the step size, and this method is called *constant-α Monte Carlo*. While Monte Carlo waits to get G_t, TD only needs to wait until next time step. At time $t + 1$, TD learning immediately forms a target and makes an update by using the observed reward R_{t+1} and the estimate $V(S_{t+1})$. The simplest TD(0) is therefore

$$V(S_t) \leftarrow V(S_t) + \alpha \left[R_{t+1} + \gamma V(S_{t+1}) - V(S_t) \right] \tag{5.32}$$

Please note the difference: The target for Monte Carlo to update is G_t, but the target for TD learning to update is $R_{t+1} + \gamma V(S_{t+1})$. Because TD learning uses its update in part on an existing estimate, it is a case of bootstrapping methods.

$$v_\pi(s) \triangleq \mathbb{E}_\pi \left[G_t \mid S_t = s \right] \tag{5.33}$$

$$= \mathbb{E}_\pi \left[\sum_{k=1}^\infty \gamma^k R_{t+k+1} \mid S_t = s \right] \tag{5.34}$$

$$=\mathbb{E}_\pi \left[R_{t+1} + \gamma \sum_{k=1}^\infty \gamma^k R_{t+k+1} \mid S_t = s \right] \tag{5.35}$$

$$=\mathbb{E}_\pi \left[R_{t+1} + \gamma v_\pi(S_{k+1}) \mid S_t = s \right] \tag{5.36}$$

TD error, which is the error in the estimation made at that time, can be defined as

$$\delta_t \triangleq R_{t+1} + \gamma V(S_{t+1}) - V(S_t) \tag{5.37}$$

where Monte Carlo error can be re-written as a series sum of TD errors.

Remark (Comparisons): TD methods learn their estimates partially on the basis of other estimates. That is, they learn a guess from a guess, bootstrap!

- TD learning has an obvious advantage over DP due to no need for a model of the environment, reward, nor next-state probability distribution or transmission.
- TD is naturally implemented in an online fully incremental fashion, but Monte Carlo must wait until the end of an episode to obtain the return.
- An important question remains untouched, can TD guarantee convergence to the correct answer? Without proof, for any fixed policy π, TD converges to v_π.
- While both TD and Monte Carlo methods converge asymptotically to the correct answer, TD is usually converging faster than constant-α Monte Carlo but general conclusion remains open.

Temporal difference methods uses raw experience which is similar to Monte Carlo methods. In the mean time, it uses part of the other learned estimates to estimate state value as well, i.e. they use value functions to refine their next prediction. This kind of methods is called *bootstrapping*. Figure 5.14 is a backup diagram to show the length of update horizon for Monte Carlo and n-step TD. Descendant states after current state S_t will serve as the realizations of state's transition and rewards. Subfigure (a) shows the back up diagram of n-step TD methods. State value of the last state, $v(S_{t+n})$, becomes an prediction for future returns after time $t + n$ (bootstrapping). Subfigure (b) is the backup diagram of Monte Carlo methods, which uses all rewards along to the terminal states.

There are various ways to incorporate rewards and state-value (or action-value) into n-step TD. A commonly applied method is the n-step tree backup: From time t to $t + n$, not only rewards $(R_{t+1}, \ldots, R_{t+n})$ but also other value function of actions that were not chosen ($Q(s_t, a), \forall a \neq A_t$) will be used to update estimation.

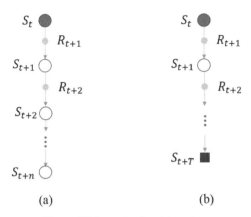

Figure 5.14   n-step bootstrapping.

The algorithm (or pseudo code) for n-step TD prediction is given in the following. Note that although there are various ways to incorporate rewards and state-value (or action-value) into n-step TD, we use the notation $G_{t:t+n}$ to summarize it.

(i) Arbitrarily initialize $V(s)$ for all $s \in \mathcal{S}$ and $a \in \mathcal{A}$. (If s is terminal state, its value must be zero.)

(ii) Initialize and store S_0 which cannot be terminal state. Input $T \leftarrow \infty$ and $t \leftarrow 0$.

(iii) If $t < T$, continue to step (iv). If not, go to step (v).

(iv) Take an action a according to policy $\pi(\cdot|S_t)$. Observe and store the next reward as R_{t+1} and the next state as S_{t+1}. If S_{t+1} is terminal, then $T \leftarrow t+1$

(v) $\tau \leftarrow t - n + 1$. ($\tau$ is the time whose state value will be updated.) If $\tau \geq 0$ go to step (vi), and go to step (ix) otherwise.

(vi) Compute the collected rewards after time τ to n steps later unless a terminal state is reached in advance.

$$G \leftarrow \sum_{i=\tau+1}^{\min(\tau+n,T)} \gamma^{i-\tau-1} R_i$$

If $\tau + n < T$, go to step (vii). Otherwise skip step (vii) and go to step (viii).

(vii) Include the estimate value of last state into the returns.

$$G \leftarrow G + \gamma^n V(S_{\tau+n})$$

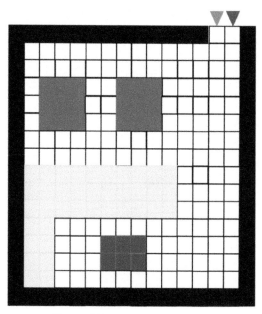

Figure 5.15 Layout of an office.

(viii) Update prediction value of state S_τ.

$$V(S_\tau) \leftarrow V(S_\tau) + \alpha[G - V(S(S_\tau))]$$

(ix) If $\tau = T - 1$, terminate; if not, let $t \leftarrow t + 1$ and goes to step ().

■ **Exercise:** A robot is going to clean an office with layout as Figure 5.15. However, this robot has no prior knowledge of office layout. The robot equips sensors to perfectly sense the status of 4 immediate neighboring grids (left, right, up, down) and move accordingly. The black grids are prohibited to enter (i.e. wall). The robot has memory to store the grids being visited and being sensed, with relative position to the entry point to form its own reference (i.e. map).

(a) Suppose all colored grids are not accessible. Please develop the RL algorithm to visit all white grids and leave from the entrance (indicated by green and red triangles) of the office. Can RL guarantee cleaning all grids? If so, how many steps does the robot take? If not, can you develop a nearly trivial method to ensure cleaning all grids?

(b) Is TD-n learning good in this case, why?

5.3.3 SARSA

Now, we turn our attention to use TD prediction methods for the control problem, a generalized policy interaction but using TD methods for the evaluation or prediction part. As applying Monte Carlo methods, we face the tradeoff between exploration and exploitation, with two major classes of on-policy and off-policy. The first step is to learn an action-value function rather than a state-value function. For an on-policy method, we must estimate $q_\pi(s, a)$ for the current policy $\pi, \forall s \in \mathcal{S}, a \in \mathcal{A}$, by using pretty much the same TD method for learning V_π while an episode consisting of an alternating sequence of states and state-action pairs.

Instead considering transitions from state to state to learn the values of states earlier, we now consider transition from state-action pair to state-action pair to learn the values of state-action pairs. Same procedure due to Markov chain with a reward process,

$$Q(S_t, A_t) \leftarrow Q(S_t, A_t) + \alpha\left[R_{t+1} + \gamma Q(S_{t+1}, A_{t+1}) - Q(S_t, A_t)\right] \quad (5.38)$$

Which gives the name of *Sarsa* algorithm due to using $(S_t, A_t, R_{t+1}, S_{t+1}, A_{t+1})$.

As in all on-policy methods, the on-policy control algorithm base on Sarsa prediction continually estimates q_π for the behavior policy π, and at the same time change π toward greediness with respect to q_π.

On-policy TD control:

(i) initialization

(ii) Choose A from S using policy derived from Q (e.g. ϵ=greedy)

(iii) For each step of the episode, until terminal state, (a) take action A and observe R and S'; (b) choose A' from S' using policy derived from Q; (c) update according to (5.38); (d) $S \leftarrow S'$ and $A \leftarrow A'$.

5.3.4 Relationship Between Q-Learning and TD-Learning

In case, we look TD learning from MDP, apart from updating value function with respect to episodes, TD learning makes immediate update at each epoch. The estimation of value function is refined by

$$v(x_k) \leftarrow v(x_k) + \alpha[R_{k+1} + \gamma v(x_{k+1}) - v(x_k)]$$

The advantages of TD over MC are efficiency especially in some applications that have long episodes.

From the view of control, the widely known *Q-learning* is indeed one of the off-policy TD control algorithm.

Q-learning (off-policy TD control):

(i) Arbitrarily initialize $Q(x, a)$ for all $x \in \mathcal{S}$ and $a \in \mathcal{A}$. (If x is terminal state, its value must be zero.)

(ii) Choose an action a in current state x based on current Q-value (i.e. action-value) estimates, e.g. ϵ-soft policies.

(iii) Take action a, then observe reward r and next state x'.

(iv) Update Q-value following

$$Q(x, a) \leftarrow Q(x, a) + \alpha[r + \gamma \max_{a'} Q(x', a') - Q(x, a)]$$

Note that $0 < \alpha \le 1$ is the learning rate. A higher value means that the refinement of Q-value occurs quickly and hence the learning goes faster. γ denotes the discount factor, also set between 0 and 1, implies that future rewards are worth less than immediate rewards.

There are two representative algorithms of TD learning, i.e. the Sarsa and the Q-learning. Sarsa stands for the "state-action-reward-state-action" and it interacts with the environment and updates the state-action value function, i.e. Q-function, based on the action it takes, while Q-learning updates the Q-function relying on the maximum reward yielded by one of available actions. Specifically, the update of the Q-function in Sarsa can be formulated in the form of:

$$Q(s, a) \leftarrow (1 - \alpha)Q(s, a) + \alpha \left[r + \gamma Q(s', a') \right], \qquad (5.39)$$

while in Q-learning, the update of the Q-function can be given by:

$$Q(s, a) \leftarrow (1 - \alpha)Q(s, a) + \alpha \left[r + \gamma \max_{a' \in \mathbb{A}} Q(s', a') \right], \qquad (5.40)$$

where s represents the system's state and a is the action selected by the agent, whilst $\mathbb{A}$ represents the available action set. Moreover, α is the update weight coefficient and γ denotes the discount factor.

Hence, Sarsa is an on-policy method[1] and yet the Q-learning is called as an off-policy method[2]. As for the convergence analysis, Sarsa is capable of converging with probability 1 to an optimal policy as well as state-action value function if all the state-action pairs are visited a large number of times. However, because of the independence of the policy of making an action

---

[1] An on-policy method learns the value of the policy that is carried out by the agent.

[2] In off-policy method, the policy used to generate behaviour is unrelated to the policy that is evaluated and improved.

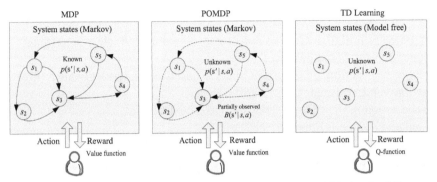

Figure 5.16   The comparison between the MDP, POMDP and TD leaning [4].

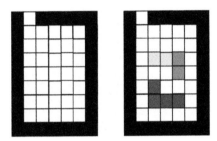

Figure 5.17   Room layout.

and that of updating the Q-function, Q-learning enables early convergence in contrast to the Sarsa.

▶**Exercise:** A robot enters a room to clean as the left of Figure 5.17 with 36 square tiles. It is obvious that the robot must enter and leave the room through the only entrance. To complete the room cleaning, the robot must go through each tile at least once. This robot can move up, down, right, and left from one tile to another neighboring, but the black tiles are prohibited to enter (i.e. wall). Without knowing the room tile layout in advance,

(a) Please use programming to find the minimum number of movements to go complete the cleaning job and to leave the room.

(b) As the right configuration of Figure 5.17, the color tiles are prohibited to move. Please use programming to find the minimum number of movements to go complete the cleaning job and to leave the room.

(c) Please develop a reinforcement learning algorithm for (b) and compare the number of movements to that in (b).

(d) Without knowing in advance, once the robot moves to yellow tiles, it will be blowed downward one tile (and thus must try to come back to clean again). Similarly, upward one tile in blue area; rightward one tile in red area; leftward one tile in green area. Please develop a reinforcement learning algorithm to complete the cleaning, and compare the results with (b).

Further Reading: [2] supplies great introduction to reinforcement learning, while the actor-critic learning is not introduced in this chapter but readers might want to have more understanding. Section 5.2.3 is taken from [3].

References

[1] M.L. Puterman, Markov Decision Processes, Wiley, 1994.
[2] R.S. Sutton, A.G. Barto, Reinforcement Learning: An Introduction, MIT Press, 1998. (2nd edition expected in October 2018 and preliminary version available online)
[3] E. Ko, K.-C. Chen, "Wireless Communications Meets Artificial Intelligence: An Illustration by Autonomous Vehicles on Manhattan Streets", *IEEE Globecom*, Abu Dhabi, 2018.
[4] J. Wang, C. Jiang, H. Zhang, Y. Ren, K.-C. Chen, Lajos Hanzo, "Thirty Years of Machine Learning: The Road to Pareto-Optimal Wireless Networks", *IEEE Communications Surveys and Tutorials*, 2020.

6

State Estimation

In order to interact with the environment (or the world), a robot can be treated as a dynamic system of internal states. Please recall from Chapter 1, the robot usually relies on sensors to acquire information about the environment. However, the sensors can be noisy in measurement and/or in transmission. Therefore, the robot must form internal belief for the states of its environment, to react through actuators. The estimation of a state for robot's operation is consequently a fundamental technology in robotics and AI.

Figure 6.1 illustrates an the interaction model for a delivery robot to use a vision sensor to develop the belief of the world. The robot must translate vision sensor data (i.e. the photo on the right) into the the belief of the world (i.e. abstraction on the left) to take appropriate actions, in order to execute tasks of navigation and delivery items by recognizing a right place.

6.1 Fundamentals of Estimation

Statistical inference allows a robot to understand the environments and effectively takes predictive actions. There are two major categories of problems in statistical inference: hypothesis testing (i.e. decision) and estimation. Unlike hypothesis testing to make a decision about parameter, estimation is to determine the actual value of parameter from observations, as accurately as possible.

Please recall the basic mathematical framework: $Y \in \mathcal{Y}, Y \sim P \in \mathcal{P}$, where $\mathcal{P} = \{P_\theta : \theta \in \Theta\}$ and $\mathcal{Y}$ is the observation space. Under this parametric model, we intend to find a function $\hat{\theta}(X)$ close to (in some sense) the unknown θ. A particular case of interest is to estimate a parameter θ from the noisy observations (embedded in additive noise w.:

$$y_i = s_i(\theta) + w_i, \quad i = 1, \ldots, n \tag{6.1}$$

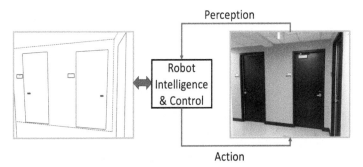

Figure 6.1   Interaction model of a robot and the environment.

There are generally two classes of estimation problems:

(a) Non-Bayesian: The parameter θ is deterministic but unknown quantity. We can derive the p.d.f of observations, $p(y \mid \theta$, called *likelihood function*. It is desirable to derive the estimator that maximizes the corresponding likelihood, called *maximum likelihood estimator* (MLE), $\hat{\theta}_{ML}$.

(b) Bayesian: θ is a (multivariate for multiple parameters) random variable, where the *a priori* p.d.f. of parameter $\pi(\theta)$ is known or available before estimation. By Bayes' rule

$$p(\theta \mid y) = \frac{p(y \mid \theta)\pi(\theta)}{p(y)} \tag{6.2}$$

The value of θ that maximizes $p(\theta \mid y)$ is called *maximum a posteriori* (MAP) estimator, $\hat{\theta}_{MAP}$.

6.1.1 Linear Estimator from Observations

For the purpose of easy computation, it is common to consider a linear estimator. We start from the relationship between two random variables.

Proposition 1. *Two random variables, Y and Z, have a joint distribution, with means m_Y, m_Z, finite variances σ_Y^2, σ_Z^2, respectively. ρ is the correlation coefficient between Y and Z. To $\mathbb{E}(Z \mid Y)$ can be represented as a linear function of Y, we have*

$$\mathbb{E}(Z \mid Y) = m_Z + \rho\frac{\sigma_Z}{\sigma_Y}(Y - m_Y) \tag{6.3}$$

$$\mathbb{E}\left[Var(Z \mid Y)\right] = \sigma_Z^2(1 - \rho^2) \tag{6.4}$$

▶ **Exercise:** Please prove above Proposition.

A general linear estimation is based on the multiple related observations. That is, to estimate a random variable Z from a number of related random variables $Y_1, \ldots, Y_N$ of known stationary statistics can be written as

$$\hat{Z} = \sum_{n=1}^{N} a_n Y_n = \mathbf{a}^T \mathbf{Y} + b \tag{6.5}$$

where $\mathbf{a} = (a_1, \ldots, a_N)$ denotes the vector of weighting coefficients that can be implemented as a linear filter; b represents the bias; and $\mathbf{Y} = (Y_1, \ldots, Y_N)$ is the vector of observations.

A common criterion to obtain the estimator is to minimize the *mean squared error* (MSE), which suggests

$$\epsilon_{LMS}^2 = \mathbb{E}\{|Z - \hat{Z}|^2\} \tag{6.6}$$

The necessary condition of minimization suggests $\mathbb{E} \mid Z - \hat{Z} \mid= 0$, which gives

$$b = m_Z - \mathbf{a}^T \mathbf{m_Y} \tag{6.7}$$

$$\hat{Z} = \mathbf{a}^T (\mathbf{Y} - \mathbf{m_Y}) + m_Z \tag{6.8}$$

Consequently,

$$
\begin{aligned}
\epsilon_{LMS}^2 &= \mathbb{E}\{|(Z - m_z) - \mathbf{a}^T (\mathbf{Y} - \mathbf{m_Y})|^2\} \\
&= \mathbb{E}\{|Z - m_Z|^2 - (Z - m_Z)(\mathbf{Y} - \mathbf{m_Y})^T \mathbf{a} \\
&\quad - \mathbf{a}(\mathbf{Y} - \mathbf{m_Y})(Z - m_Z) + \mathbf{a}^T (\mathbf{Y} - \mathbf{m_Y})(\mathbf{Y} - \mathbf{m_Y})^T \mathbf{a}\}
\end{aligned}
$$

By defining covariance matrices

$$\mathbf{C_{YZ}} = \mathbb{E}\{(\mathbf{Y} - \mathbf{m_Y})(Z - m_Z)\} \tag{6.9}$$

$$\mathbf{C_Y} = \mathbb{E}\{(\mathbf{Y} - \mathbf{m_Y})(\mathbf{Y} - \mathbf{m_Y})^T\} \tag{6.10}$$

we can rewrite the MSE as

$$\epsilon_{LMS}^2 = \sigma_Z^2 - \mathbf{C_{YZ}^T} \mathbf{a} - \mathbf{a} \mathbf{C_{YZ}} + \mathbf{a}^T \mathbf{C_Y} \mathbf{a} \tag{6.11}$$

The necessary condition to retain the least MSE is

$$\nabla_{\mathbf{a}} \epsilon_{LMS}^2 = 0 \tag{6.12}$$

which gives $\mathbf{C_Y} \mathbf{a} = \mathbf{C_{YZ}}$.

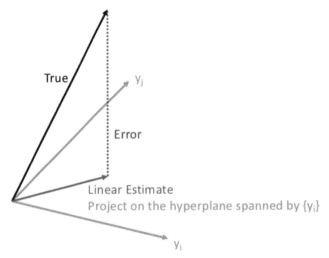

Figure 6.2   Linear estimator on the hyperplane spanned by observations orthogonal to the error.

Proposition 2. *The coefficients of linear estimation on the multiple correlated observations are*

$$\mathbf{a} = \mathbf{C_Y}^{-1}\mathbf{C_{Y}}_Z \tag{6.13}$$

while the resulting MSE is

$$\epsilon_{LMS}^2 = \sigma_Z^2 - \mathbf{C}_{YZ}^T\mathbf{C_Y}^{-1}\mathbf{C}_{YZ} \tag{6.14}$$

Remark (Orthogonal Principle): The engineering meaning of linear estimation (i.e. linear combination of observations to obtain the estimator) can be intuitively and geometrically illustrated by Figure 6.2. The observations span a hyperplane and the linear estimator is actually the projection of true value vector onto the hyperplane. The error is the difference between the true vector and linear estimator, and thus orthogonal to linear estimator. This property is useful for derivations and calculations, and known as *orthogonal principle* or *projection theorem*.

▶ **Exercise:** We intend to the linear estimate of a random signal A from noisy observations $x_n, n = 0, \ldots, N - 1$ (that is, the estimate is a linear combination of x_n). Suppose the signal model to be

$$x_n = A + w_n \tag{6.15}$$

where $A \sim \mathcal{U}[-A_0, A_0]$, w_n is white Gaussian noise with zero mean and variance σ^2, and A and w_n are independent. Please find $\hat{A}$.

6.1.2 Linear Prediction

As indicated in earlier chapter, the linear prediction of a stationary random process can be realized with the assistance of Wiener-Hopf equation. Or, the linear prediction can take advantage of linear regression over a training dataset.

Example: Suppose the behavior of an unknown system can be observed by the input variables $\mathbf{X} = (X_1, \ldots, X_r)$ and output variable Y, such that we can construct the following linear model to predict the output of Y.

$$\hat{Y} = \hat{b}_0 + \sum_{j=1}^{r} \hat{b}_j X_j \qquad (6.16)$$

where $\hat{b}_0$ is known as the *bias*, which usually can be included in $\mathbf{X}$. If denote $\hat{\mathbf{b}} = (\hat{b}_1, \ldots, \hat{b}_r)$, then

$$\hat{Y} = \mathbf{X}^T \hat{\mathbf{b}} \qquad (6.17)$$

The most common performance measure for prediction is *least square-error*. In machine learning, such an approach is to identify coefficients b to minimize the *residual sum of squares*. For N consecutive observations on $(\mathbf{X}_i, Y_i)$, $i = 1, \ldots, N$, which can be viewed as the training set, then

$$RSS(\mathbf{b}) = \sum_{i=1}^{N} (y_i - \mathbf{x}_i^T \mathbf{b})^2 = (\mathbf{y} - \mathbf{X}\mathbf{b})^T (\mathbf{y} - \mathbf{X}\mathbf{b}) \qquad (6.18)$$

$RSS(\mathbf{b})$ is a quadratic function of parameters and hence minimum always exists but may not be unique. Again, the necessary condition of optimality gives normal equations

$$\hat{\mathbf{b}} = (\mathbf{X}^T \mathbf{X})^{-1} \mathbf{X}^T \mathbf{y} \qquad (6.19)$$

Please note that above equation is precisely as the linear regression, which allows linear prediction of a unknown system.

Lemma (Ridge Regression): *Ridge regression* shrinks the regression coefficients by imposing a penalty on their size. The ridge coefficients minimize a penalized RSS.

$$\hat{b}^{ridge} = \underset{b}{\mathrm{argmin}}\{\sum_{i=1}^{N}\left(y_i - b_0 - \sum_{j=1}^{r} x_{ij}b_j\right)^2 + \lambda\sum_{j=1}^{r} b_j^2\} \qquad (6.20)$$

Lemma (LASSO): The *least absolute shrinkage and selection operator* (LASSO) method is evolved from Ridge regression using L_p norm.

$$\hat{b}^{lasso} = \underset{b}{\mathrm{argmin}}\{\sum_{i=1}^{N}\left(y_i - b_0 - \sum_{j=1}^{r} x_{ij}b_j\right)^2 + \lambda\sum_{j=1}^{r} \|b_j\|_p\} \qquad (6.21)$$

which is equivalent to ridge regression if $p = 2$. LASSO has been widely applied in statistical inference and data analysis to manipulate the convergence behavior (such as the speed of convergence) and overfitting.

6.1.3 Bayesian Estimation

For a random observation Y, a family of distributions indexed by a parameter $\theta \in \Theta \subseteq \mathbb{R}$. The goal of parameter estimation is to find a function $\hat{\theta}(y)$ to serve the best guess of the true value of θ based on $Y = y$. The cost function $C[a, \theta]$ is the cost estimating a true value of θ as $a \in \Theta, \forall \theta$. The conditional risk averaged over $Y, \forall\Theta$ is

$$r_\theta(\hat{\theta}) = \mathbb{E}_\theta\{C\left[\hat{\theta}(Y), \theta\right]\} \qquad (6.22)$$

Similar to hypothesis testing, the average Bayesian risk is defined as

$$R(\hat{\theta}) = \mathbb{E}\{r_\theta(\hat{\theta})\} \qquad (6.23)$$

The goal is to device an estimator minimizing $R(\hat{\theta})$, which is known as the *Bayesian estimate* of θ. Based on different definitions of the cost function, there are three common classes of Bayesian estimation.

- The *mean squared error* (MSE) is most common to establish the cost as

$$C[a, \theta] = (a - \theta)^2 \qquad (6.24)$$

The consequent Bayesian risk is $\mathbb{E}\{\left[\hat{\theta}(Y) - \Theta\right]^2\}$. The resulting Bayesian estimate is named as *minimum mean squared error estimator*

(MMSE). The posterior risk given $Y = y$ is $\mathbb{E}\{\left[\hat{\theta}(Y) - \Theta\right]^2 \mid Y = y\}$. Differentiating to obtain the necessary condition of optimality, the Bayesian estimate

$$\hat{\theta}_{MMSE}(y) = \mathbb{E}\{\Theta \mid Y = y\} \tag{6.25}$$

which is the conditional mean of Θ given $Y = y$.
- Another cost function is defined as

$$C[a, \theta] = |a - \theta| \tag{6.26}$$

which yields the Bayesian risk $\mathbb{E}\{|\hat{\theta}(Y) - \Theta|\}$ and thus the *minimum mean absolute error estimate* (MMAE), $\hat{\theta}_{ABS}(y)$. Please note that $\hat{\theta}_{ABS}(y)$ is actually the *median* of the conditional distribution of Θ given $Y = y$.
- For a cost function very similar to uniform cost as follows

$$C[a, \theta] = \begin{cases} 0, & if \ |a - \theta| \leq \Delta \\ 1, & if \ |a - \theta| > \Delta \end{cases} \tag{6.27}$$

The subsequent Bayesian estimate $\hat{\theta}_{MAP}$ represents the *maximum a posteriori probability* (MAP) of Θ given $Y = y$. Suppose p_θ to be the probability density function given $\Theta = \theta$ and $w(\theta)$ is the prior distribution. $\hat{\theta}_{MAP}$ is obtained by maximizing $p_\theta(y)w(\theta)$.

▶ **Exercise:** Let the observations

$$r_i = a + n_i, \quad i = 1, \ldots, N \tag{6.28}$$

where $a \sim G(0, \sigma_a^2)$, and $n_i \sim G(0, \sigma_n^2)$ are independent. Please show that

$$\hat{a}_{MMSE} = \frac{\sigma_a^2}{\sigma_a^2 + (\sigma_n^2/N)} \left(\frac{1}{N} \sum_{i=1}^{N} r_i \right) \tag{6.29}$$

Please also find $\hat{a}_{ABS}$ and $\hat{a}_{MAP}$.

Definition: An estimate is called *unbiased* if $\mathbb{E}_\theta\{\hat{\theta}(Y)\} = \theta$.

When the conditional MSE serves as the variance of the estimate, an unbiased estimate minimizing MSE is called a *minimum variance unbiased estimator* (MVUE).

▶ **Exercise:** Observation data $x_n, n = 1, \ldots, N$ are independently and identically distributed as $G \sim (0, \sigma^2)$. We wish to estimate the variance σ^2 as

$$\hat{\sigma}^2 = \frac{1}{N} \sum_{n=1}^{N} x_n^2 \tag{6.30}$$

Is this an unbiased estimator? Find the variance of $\hat{\sigma}^2$ and examine the situation when $N \to \infty$.

Definition: *Fisher's information I_θ is*

$$I_\theta = \mathbb{E}_\theta\{(\frac{\partial}{\partial \theta} \log p_\theta(Y))^2\} \tag{6.31}$$

which can be also computed as

$$I_\theta = -\mathbb{E}_\theta\{\frac{\partial^2}{\partial \theta^2} \log p_\theta(Y)\} \tag{6.32}$$

Proposition (Cramer-Rao Lower Bound): In case $\hat{\theta}$ is unbiased,

$$Var_\theta\left[\hat{\theta}(Y)\right] \geq \frac{1}{I_\theta} \tag{6.33}$$

▶ **Exercise:** A single sample is observed as

$$x_1 = A + w_1 \tag{6.34}$$

where $w_1 \sim G(0, \sigma^2)$. Please find the variance of estimator $\hat{A}$. Is it consistent with the Cramer-Rao Low bound?

▶ **Exercise:** Following the previous problem, instead, we use multiple independent observations to estimate A, by

$$x_n = A + w_n, n = 1, \ldots, N \tag{6.35}$$

where $w_n \sim G(0, \sigma^2)$. Please find $\hat{A}$ and the Cramer-Rao lower bound of $\hat{A}$.

▶ **Exercise:** Following previous problem, however, $x_n, n = 1, \ldots, N$ are not white anymore (i.e. color noise), with covariance matrix $\mathbf{Q}$ formed by $\mathbf{x} = (x_1, \ldots, x_N)$. Please find the minimum variance estimator of A.

▶ **Exercise:** Similar to previous problems, we observe

$$x_n = A + w_n, n = 1, \ldots, N \tag{6.36}$$

However, w_n are zero mean and uncorrelated noise with variance σ_n^2. Please find the best linear unbiased estimator (BLUE) of A.

6.1.4 Maximum Likelihood Estimation

In practice, it may not be feasible to find MVUEs. A common alternative is maximum-likelihood method, which was first suggested by C.F. Gauss in 1821. However, this approach is usually credited to R.A. Fisher in 1922, who rediscovered the idea.

Please recall the MAP estimation

$$\hat{\theta}_{MAP}(y) = \arg\{\max_a p_\theta(y)w(\theta)\} \tag{6.37}$$

In absence of any prior information about the parameter, we usually assume uniform distribution for the parameter, as it represents more or less a worst-case scenario, to complete the estimation. Since $p_\theta(y)$ is known as the *likelihood* function,

$$\hat{\theta}_{ML} = \arg\{\max_a p_\theta(y)\} \tag{6.38}$$

is known as the *maximum likelihood estimate* (MLE).

▶ **Exercise:** If we observe N i.i.d. samples from a Bernoulli experiment with probabilities

$$P\{x_n = 1\} = p = 1 - P\{x_n = 0\}, 0 < p < 1 \tag{6.39}$$

Please find the maximum likelihood estimator of p.

▶ **Exercise:** The time T between requests for services in a system is distributed as

$$f_T(T) = \alpha e^{-\alpha T}, \ T \geq 0 \tag{6.40}$$

where $\alpha > 0$ is a constant known as the arrival rate. Totally $N + 1$ requests have been observed with independent inter-arrival times $T_1, \ldots, T_N$. Please find the estimate of α.

MLE has wide applications in robotics. The environment or the state of the agent can be described by a set of parameters, and MLE serves as a useful methodology of parameter estimation.

Example (Location Estimation in a Sensor Network): To determine the position of a robot, which is known as *localization* in later chapter, we may use a network of sensors to estimate the location. As Figure 6.3 with location being simplified as one-dimensional parameter θ, the measurements from this set of N sensors are $x_n = \theta + w_n$ and corrupted by an additive noise w_n, $n = 1, \ldots, N$. Suppose the measurements x_n are transmitted to the fusion

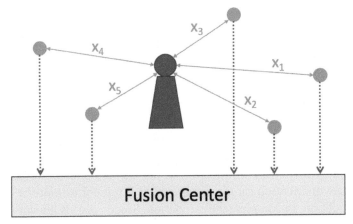

Figure 6.3 Sensors ($N = 5$) measure distances to the robot and transmit to the fusion center to conclude robot's location.

center, under the criterion of mean squared error (MSE, $\mathbb{E}\{\|\theta - \hat{\theta}\|^2\}$), the maximum likelihood estimator (MLE)

$$\hat{\theta} = \frac{1}{N} \sum_{n=1}^{N} x_n \tag{6.41}$$

which is actually the em sample mean of the measurements, and *unbiased* (i.e. $\mathbb{E}\hat{\theta} = \theta$). In case of Gaussian noise $w_n \sim G(0, \sigma^2)$, $Var(\hat{\theta}) = \frac{\sigma^2}{N}$. Such a method to estimate location is generally known as distributed estimation in wireless sensor networks.

Example (Bandwidth-Efficient Location Estimation in a Sensor Network): The way to obtain location estimation as Figure 6.3 is straightforward but requires totally N times of wireless communication to transfer x_n. Considering the multiple access (details in Chapter 10), location estimation can take a good amount of radio resource (i.e. bandwidth). Therefore, bandwidth-efficient (location) estimation is consequently desirable [3]. Instead of transmitting x_n, we can compress x_n into $m_n(x_n)$, with minimum of one-bit data, that is, $m_n(x_n) = 0$ or 1.

Again, suppose $w_n \sim G(0, \sigma^2)$, then

$$m_n(x_n) = m_n = \begin{cases} 1, & x_n \geq \eta \\ 0, & x_n < \eta \end{cases} \tag{6.42}$$

where η is a threshold parameter to supply the information of approximate location of θ.

$$\mathbb{P}\{m_n(x_n) = 1\} = Q(\eta - \theta) \tag{6.43}$$

where $Q(x) = \frac{1}{\sqrt{2\pi}} \int_x^\infty e^{-t^2/2} dt$, also known as *Gaussian tail function*. At the fusion center, Due to independent measurements and noises, the collected sensor data vector $\mathbf{m} = (m_1, \ldots, m_N)$ is distributed in Bernoulli with parameter $Q(\eta - \theta)$. The likelihood function is

$$p(\mathbf{m}, \theta) = \prod_{n=1}^{N} [Q(\eta - \theta)]^{m_n} [1 - Q(\eta - \theta)]^{1-m_n} \tag{6.44}$$

Using the log-likelihood and the necessary condition of optimum, the MLE is

$$\hat{\theta} = \eta - Q^{-1} \left[\frac{1}{N} \sum_{n=1}^{N} m_n(x_n) \right] \tag{6.45}$$

Please note that the optimal selection of $\eta = \theta$ gives the variance $\frac{\pi \sigma^2}{2N}$, which is just $\pi/2$ times higher than earlier example by just transmitting one-bit data from each sensor instead of real-value.

▶ **Exercise:** Phase information is critical in some techniques for a robot to explore the environment. For example, distance information can be derived from the phase of a reflected wave, which serves the principle of mmWave warning radar when backing off a car. Suppose we intend to estimate the unknown constant phase θ of a sinusoid waveform embedded in additive white Gaussian noise (with zero mean and 2-side p.s.d. $N_0/2$)

$$x(t) = \sqrt{2}A \cos(2\pi f_c t + \theta) + w(t) \tag{6.46}$$

Please find the $\hat{\theta}$.

6.2 Recursive State Estimation

To deal with the dynamic systems, instead single-shot estimation, recursive estimation is preferred in many cases of robotics as its special feature. Again, the environment is characterized by the *state*, which can be viewed as the collection of all aspects of the robot and corresponding environment. State can be static or changing over time, while we usually denote x_t as the state at time t. The state may be robot pose, velocity, location, features of environment, etc. A state x_t is called *complete* if it is the best predictor of the future. Completeness entails the knowledge of past states, measurements, etc. to help predicting the future. Once the future being stochastic, *Markov chain* is commonly applied for this temporal purpose. In Figure 6.1, the robot interacting with environment involves sensor measurements, z_t and control actions u_t.

To deal with uncertainty in the future, the evolution of state and measurements is governed by probabilistic laws, and thus the state can be described as

$$p(x_t \mid x_{0:t-1}, z_{1:t-1}, u_{1:t}) \tag{6.47}$$

If the state x is complete to sufficiently summarize all happening before, then

$$p(x_t \mid x_{0:t-1}, z_{1:t-1}, u_{1:t}) = p(x_t \mid x_{t-1}, u_t) \tag{6.48}$$

which is the *state transition probability* in a Markov process. Similarly,

$$p(z_t \mid x_{0:t-1}, z_{1:t-1}, u_{1:t}) = p(z_t \mid x_t) \tag{6.49}$$

which is known as the *measurement probability*.

Please note that the global view on the interaction between a robot and the environment can be different from the local view of a robot itself. Therefore, the internal knowledge of a robot regarding the state of environment is called *belief*, which serves as an important concept of *probabilistic robotics*. For example, in Figure 6.1, a delivery that is actually in coordinate (l_x, l_y, l_z, l_t) robot must infer its location as GPS is neither available nor precise enough. To distinguish the true state, inference by this robot is known as belief or state of knowledge.

The belief of the state x_t can be denoted as

$$\mathbb{B}(x_t) = p(x_t \mid z_{1:t}, u_{1:t}) \tag{6.50}$$

Sometimes, it is useful to compute *a posterior* prior to incorporating z_t and right after the control u_t. Such a posterior is denoted as

$$\bar{\mathbb{B}}(x_t) = p(x_t \mid z_{1:t-1}, u_{1:t}) \tag{6.51}$$

which is often referred as *prediction*. Then, computing $\mathbb{B}(x_t)$ from $\bar{\mathbb{B}}(x_t)$ is called *correction* or *measurement update*.

The belief of the system states actually indicates a particular nature in dynamic systems, known as a *hidden process* as shown in Figure 6.4.

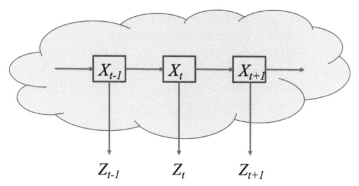

Figure 6.4 The dynamic system has states X_t that are hidden (shown in the cloud-shade area), and the observation process delivers Z_t that can be viewed as measurements.

If this dynamic system cab be represented as a Markov process, $p(X_t \mid X_{t-1}, \ldots, X_1) = p(X_t \mid X_{t-1})$, the *hidden Markov model* (HMM) can be applied. HMM has been successfully applied in many engineering applications such as speech recognition [5] and recursive estimation of robot's locaiton later in this chapter.

Remark: In statistical learning, for each observation Z_t, we view X_t as latent variables forming a Markov chain. In the following, the latent variables are discrete to form the HMM, but the observations can be continuous (in theory, can be discrete too) due to the interest in robotics. When dealing with linear systems, the latent variables and observations are usually Gaussian, by recalling the fact that Gaussian input of a linear system results in jointly Gaussian output.

Denote $\mathcal{X} = \{x_1, \ldots, x_N\}$ as a set of N system states; $\mathbb{P} = [p_{ij}]$ as the state transition probability matrix; $Z : z_1, \ldots, z_T$ as a sequence of T observations; $\mathcal{E} = \eta_i(z_i)$ as a sequence of observation likelihoods, also known as *emission probabilities*, which means the probability of z_i being generated from the state i; $\pi^1, \ldots, \pi^N$ as the initial probability distribution over states, that is, π^i is the probability that the Markov chain starts from state i, while $\sum_{i=1}^{N} \pi^i = 1$. In addition to system Markovian property, it is assumed that an output observation z_i depends only on the state x_i, not any other states nor observations.

Example: A well known example of HMM was given by Jason Eisner: Imagine that Clare is a climatologist in the year 2999 after a nuclear war in 2050. Almost all the records of the weather are lost, particularly the ancient city of Miami is in the ocean now. However, Clare is lucky to find a sales record of ice creams in a big mall in 2030. The problem turns out to be: Given a sequence of observations Z (each integer-valued in $z_1, \ldots, z_T$ representing the number of ice cream sales on a given day), find the hidden sequence $\mathcal{X}$ of weather states (say, hot or cool) which causes people to consume ice cream.

Generally speaking, there are three kinds of fundamental problems for HMM and subsequent algorithms to compute:

Likelihood Given an HMM $\xi = (\mathbb{P}, \mathcal{E})$ and an observation sequence Z, determine the likelihood $P(Z \mid \xi)$.

Decoding Given an observation sequence Z and an HMM $\xi = (\mathbb{P}, \mathcal{E})$, discover the most probable hidden state sequence from $\mathcal{X}$. This class of problems is equivalent to the well known *Viterbi algorithm* used in optimal decoding of convolutional codes and in the *maximum likelihood sequence estimation* of statistical communication theory.

Learning Given an observation sequence Z and the set of states $\mathcal{X}$ in HMM, learn the parameters $\mathbb{P}$ and $\mathcal{E}$ of the HMM.

Further details of computing algorithms can be found in [5].

6.3 Bayes Filters

A fundamental way to compute the belief takes advantage of *Bayes Theorem* in a recursive manner to compute $\mathbb{B}(x_t)$ from $\mathbb{B}(x_{t-1})$, which can be realized by computing a belief over the state x_t based on the prior belief over state x_{t-1} and the control u_t. That is, the belief that the robot assigns to state x_t is obtained by the integral (or summation) of the product of two distributions: the prior assigned to x_{t-1}, and the probability that control u_t induces a transition from x_{t-1} to x_t.

$$\bar{\mathbb{B}}(x_t) = \int p(x_t \mid u_t, x_{t-1}) \bar{\mathbb{B}}(x_{t-1}) dx_{t-1} \qquad (6.52)$$

This update step is known as *control update* or *prediction*.

The second step of the Bayes filter is called the measurement update. For all hypothetical posterior state x_t, the belief is updated by the measurement z_t as follows.

$$\mathbb{B}(x_t) = c \cdot p(z_t \mid x_t) \bar{\mathbb{B}}(x_t) \qquad (6.53)$$

where c is the normalization constant for total probability.

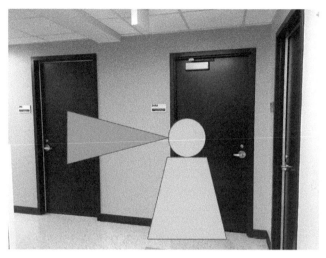

Figure 6.5   A robot to estimate whether the door is open or not.

Proposition 3 (Bayes Filter Algorithm). *The Bayes filter algorithm can be summarized as the following steps:*

1. *For all x_t, compute (6.52) and (6.53).*
2. *Return $\mathbb{B}(x_t)$.*

Example [4]: A delivery robot intends to estimate whether the door of office ENB 245 is open as Figure 6.5. To simplify the problem, we assume the door is either open or closed, two possible states, and the robot has no initial knowledge of the state for this door.

Without any prior information, the robot sets equal probability to two possible state of the door.

$$\mathbb{B}(X_0 = open) = 0.5 \tag{6.54}$$
$$\mathbb{B}(X_0 = closed) = 0.5 \tag{6.55}$$

The noisy sensor of the robot observes as the following conditional probabilities (please note the similarity to binary hypothesis testing in digital communication systems).

$$p(Z_t = open \mid X_t = open) = 0.6 \tag{6.56}$$
$$p(Z_t = closed \mid X_t = open) = 0.4 \tag{6.57}$$
$$p(Z_t = open \mid X_t = closed) = 0.2 \tag{6.58}$$
$$p(Z_t = closed \mid X_t = closed) = 0.8 \tag{6.59}$$

The robot can use a actuator to open the door. If the door is already open, it will remain open. It the door is closed, the robot will open the door afterwards with probability 0.8:

$$p(X_t = open \mid U_t = push, X_{t-1} = open) = 1 \qquad (6.60)$$

$$p(X_t = closed \mid U_t = push, X_{t-1} = open) = 0 \qquad (6.61)$$

$$p(X_t = open \mid U_t = push, X_{t-1} = closed) = 0.8 \qquad (6.62)$$

$$p(X_t = closed \mid U_t = push, X_{t-1} = closed) = 0.2 \qquad (6.63)$$

The robot can choose not to use the actuator to open the door (i.e. $\times$), which means the state (of the world) remains the same.

$$p(X_t = open \mid U_t = \times, X_{t-1} = open) = 1 \qquad (6.64)$$

$$p(X_t = closed \mid U_t = \times, X_{t-1} = open) = 0 \qquad (6.65)$$

$$p(X_t = open \mid U_t = \times, X_{t-1} = closed) = 0 \qquad (6.66)$$

$$p(X_t = closed \mid U_t = \times, X_{t-1} = closed) = 1 \qquad (6.67)$$

Suppose at time $t = 0$, the robot takes no control action but senses an open door. The resulting posterior belief is calculated by the Bayes filter using the prior belief $\mathbb{B}(x_0)$, the control $u_1 = \times$, and the measurement senses open as input. According to (6.52), we have

$$\mathbb{B}(x_1) = \sum_{x_0} p(x_1 \mid u_1, x_0)\mathbb{B}(x_0)$$

$$= p(x_1 \mid U_1 = \times, X_0 = open)\mathbb{B}(X_0 = open)$$
$$+ p(x_1 \mid U_1 = \times, X_0 = closed)\mathbb{B}(X_0 = closed)$$

Now, we are ready to compute two hypotheses of X_1:

$$\bar{\mathbb{B}}(X_1 = open) = 1 \cdot 0.5 + 0 \cdot 0.5 = 0.5 \qquad (6.68)$$
$$\bar{\mathbb{B}}(X_1 = closed) = 0 \cdot 0.5 + 1 \cdot 0.5 = 0.5 \qquad (6.69)$$

Since the action for the robot is not to act, it is not a surprise to get $\bar{\mathbb{B}}(x_1) = \mathbb{B}(X_0)$. However, by incorporating the measurement, the belief may be changed. According to (6.53),

$$\mathbb{B}(x_1) = c \cdot p(Z_1 = open \mid x_1)\bar{\mathbb{B}}(x_1) \qquad (6.70)$$

For two possibilities of the state X_1,

$$
\begin{aligned}
\mathbb{B}(X_1 = open) &= c \cdot p(Z_1 = open \mid X_1 = open)\bar{\mathbb{B}}(X_1 = open) \\
&= c(0.6)(0.5) \\
\mathbb{B}(X_1 = closed) &= c \cdot p(Z_1 = open \mid X_1 = closed)\bar{\mathbb{B}}(X_1 = closed) \\
&= c(0.2)(0.5)
\end{aligned}
$$

Recall c is a normalization factor, and $c = 2.5$. Therefore,

$$
\mathbb{B}(X_1 = open) = 0.75
$$
$$
\mathbb{B}(X_1 = closed) = 0.25
$$

▶ **Exercise:** Please continue above example to compute $\mathbb{B}(X_2)$. In case both measurements are correct, do you think this robot is reliable?

▶ **Exercise:** Please use the induction method to prove the Bayes Filter Algorithm in Proposition 1.

6.4 Gaussian Filters

Gaussian Filters represent an important family of recursive state estimators, in which the beliefs are represented by multi-variate Gaussian distributions as

$$
p(x) = \det(2\pi\Sigma)^{-\frac{1}{2}} \exp\left[-\frac{1}{2}(x - \mu)^T\Sigma^{-1}(x - \mu)\right] \tag{6.71}
$$

where μ is the mean and Σ is the covariance matrix.

6.4.1 Kalman Filter

As indicated in Chapter 3, the optimal filter for stationary signals is known as *Wiener filter*. For robotics, the operating environment can be very dynamic, unknown or without *priori* information, and non-stationary. The optimal filter dealing with non-stationary signals is *Kalman filter*. The Kalman filter was invented in the 1950's by Rudolph Emil Kalman, for the purpose of filtering and prediction in linear systems.

Different from original optimal filtering, we intend to use Kalman filter to represent beliefs by the moments: The belief is represented by the mean μ_t and covariance Σ_t at time t. Posteriors are Gaussian, in addition to Markov assumptions of Bayes filter, if the following three properties hold:

1. The probability of next state, $p(x_t \mid u_t, x_{t-1})$, must be a linear function with additive Gaussian noise.

$$x_t = A_t x_{t-1} + B_t u_t + \epsilon_t \tag{6.72}$$

where $\{x_t\}$ are state vectors and u_t is the control vector at time t.

$$x_t = \begin{pmatrix} x_{1,t} \\ x_{2,t} \\ \vdots \\ x_{n,t} \end{pmatrix} \text{ and } u_t = \begin{pmatrix} u_{1,t} \\ u_{2,t} \\ \vdots \\ u_{m,t} \end{pmatrix} \tag{6.73}$$

where A_t is a $n \times n$ matrix and B_t is a $n \times m$ matrix. By this way, the Kalman filter represents linear system dynamics. ϵ_t is a n-dimensional Gaussian random vector modeling the randomness in state transition, and has zero mean and covariance R_t. A state transition probability of the form (6.72) is called a *linear Gaussian*. Equation (6.72) defines the state transition probability $p(x_t \mid u_t, x_{t-1})$. By incorporating multivariate Gaussian distribution,

$$p(x_t \mid u_t, x_{t-1}) = \det(2\pi R_t)^{-\frac{1}{2}}$$
$$\cdot e^{-\frac{1}{2}(x_t - A_t x_{t-1} - B_t u_t)^T R_t^{-1}(x_t - A_t x_{t-1} - B_t u_t)}$$

$$\tag{6.74}$$

2. The probability of measurement $p(z_t \mid x_t)$ is also linear with additive Gaussian noise.

$$z_t = C_t x_t + n_t \tag{6.75}$$

where C_t is a $k \times n$ matrix, k is the dimension of the measurement vector z_t, and n_t is the measurement noise as a multivariate Gaussian with zero mean and covariance matrix Ψ_t.

3. The initial belief $\mathbb{B}(x_0)$ must be Gaussian with mean μ_0 and covariance Σ_0.

Proposition 4. [4] Kalman Filter Algorithm: *Input:* $\mu_{t-1}, \Sigma_{t-1}, u_t, z_t$.

$$\bar{\mu}_t = A_t \mu_{t-1} + B_t u_t \tag{6.76}$$
$$\bar{\Sigma}_t = A_t \Sigma_{t-1} A_t^T + R_t \tag{6.77}$$
$$\kappa_t = \bar{\Sigma}_t C_t^T (C_t \bar{\Sigma}_t C_t^T + Q_t)^{-1} \tag{6.78}$$

$$\mu_t = \bar{\mu}_t + \kappa_t(z_t - C_t\bar{\mu}_t) \tag{6.79}$$
$$\Sigma_t = (I - \kappa_t C_t)\bar{\Sigma}_t \tag{6.80}$$

Return: μ_t, Σ_t.

Remark: At time t, the Kalman filter represents its belief with mean μ_t and covariance Σ_t. With the input about its belief at time $t-1$, the control u_t and the measurement z_t update the Kalman filter.

The first two equations indicate calculating parameters $\bar{\mu}_t, \bar{\Sigma}_t$ of predicted belief one step later without including the measurement z_t, by substituting μ_{t-1} for the state x_{t-1} in (6.72). The update of the covariance considers the fact that states depend on previous states through the linear transformation A_t, while this matrix is multiplied twice into the covariance since the covariance is a quadratic matrix.

By taking z_t into account, the last three equations transform $\bar{\mathbb{B}}(x_t)$ to the desired belief $\mathbb{B}(x_t)$, by first computing the variable κ_t that is called *Kalman gain*. Kalman gain specifies how much the measurement is taken into account to estimate new state, and is used to online update. Finally, the new covariance of the posterior belief is adjusted for the *information gain* resulting from the measurement.

Remark: The Kalman filter can be computed in a quite efficient manner, whose complexity is pretty much dominated by the matrix inversion.

6.4.2 Scalar Kalman Filter

To derive above matrix equations of Kalman filter involves a lot of efforts. In this sub-section, based on the derivations of linear estimation in Section 6.1, the scalar *Kalman filter* that is the simplest form of Kalman filter will be derived. A scalar Kalman filter means scalar states and scalar observations. The simple scalar Gauss-Markov signal model is considered as

$$x_n = ax_{n-1} + u_n, \; n = 0, 1, \ldots \tag{6.81}$$

where $u_n \sim G(0, \sigma_u^2)$. A sequential MMSE estimator to estimate x_n is based on observations $z_0, z_1, \ldots, z_n$, and the observation model is

$$z_n = x_n + w_n \tag{6.82}$$

where $w_n \sim G(0, \sigma_n^2)$ is varying with time index n. The purpose of Kalman filter is to compute $\hat{x}_n$ based on $\hat{x}_{n-1}$. It is reasonable to assume x_{-1}, u_n, w_n are independent, and $x_{-1} \sim G(0, \sigma_x^2)$. The goal is to estimate x_n based on observations $\{z_0, z_1, \ldots, z_n\}$, or to filter z_n to yield x_n. Denote $\hat{x}_{n|m}$ as the estimator of x_n based on the observations $\{z_0, z_1, \ldots, z_m\}$. The criterion of optimality is to minimize Bayesian MSE, that is,

$$\mathbb{E}\left[(x_n - \hat{x}_{n|n})^2\right]$$

the expectation with respect to $p(z_0, \ldots, z_n; x_n)$. Applying (6.13) and joint Gaussian, the MMSE estimator is the mean of the posterior p.d.f.

$$\hat{x}_{n|n} = \mathbb{E}(x_n \mid z_{0:n}) = \mathbf{C}_{xz}\mathbf{C}_{zz}^{-1}\mathbf{z} \tag{6.83}$$

Since the signals and noise are Gaussian, the MMSE estimator in linear is equivalent to linear MMSE (LMMSE). If they are not Gaussian, this is just LMMSE.

Remark: If z_n is uncorrelated with $\{z_0, z_1, \ldots, z_{n-1}\}$, (6.83) and orthogonal principle give

$$\hat{x}_{n|n} = \mathbb{E}(x_n \mid z_{0:n-1}) + \mathbb{E}(x_n \mid z_n)$$
$$= \hat{x}_{n|n-1} + \mathbb{E}(x_n \mid z_n)$$

which is in recursive form to serve our purpose. Unfortunately, $\{z_n\}$ are correlated, but thus useful to estimate.

Lemma: Considering the sequential LMMSE, we have the properties:

(i) The MMSE estimator of θ based on two uncorrelated data vectors $\mathbf{y}_1, \mathbf{y}_2$, assuming jointly Gaussian, is

$$\hat{\theta} = \mathbb{E}(\theta \mid \mathbf{y}_1, \mathbf{y}_2) \tag{6.84}$$
$$= \mathbb{E}(\theta \mid \mathbf{y}_1) + \mathbb{E}(\theta \mid \mathbf{y}_2) \tag{6.85}$$

(ii) The MMSE estimator is additive if $\theta = \theta_1 + \theta_2$, then

$$\hat{\theta} = \mathbb{E}(\theta \mid \mathbf{y}) = \mathbb{E}(\theta_1 + \theta_2 \mid \mathbf{y}) \tag{6.86}$$
$$= \mathbb{E}(\theta_1 \mid \mathbf{y}) + \mathbb{E}(\theta_2 \mid \mathbf{y}) \tag{6.87}$$

Denote $\mathbf{Z_n} = (z_0, z_1, \ldots, z_n)^T$ and thus $\mathbf{z_n}$ is an observation (data) vector. Let $\tilde{z}_n$ denote the innovation that represents the part of z_n that is uncorrelated with previous observations $z_0, \ldots, z_{n-1}$.

$$\tilde{z}_n = z_n - \hat{z}_{n|n-1} \tag{6.88}$$

where the orthogonality suggests $\hat{z}_{n|n-1}$ is on the span $\{z_0, \ldots, z_{n-1}\}$. With the assistance of (6.13), another way to look at (6.88) is

$$z_n = \tilde{z}_n + \hat{z}_{n|n-1} \tag{6.89}$$

$$= \tilde{z}_n + \sum_{k=1}^{n-1} a_k z_k \tag{6.90}$$

(6.83) can be re-written as

$$\hat{x}_{n|n} = \mathbb{E}(x_n \mid \mathbf{Z}_{n-1}, \tilde{z}_n) \tag{6.91}$$

where $\mathbf{Z}_{n-1}$ and $\tilde{z}_n$ are uncorrelated. Then, property (i) in the Lemma gives

$$\hat{x}_{n|n} = \mathbb{E}(x_n \mid \mathbf{Z}_{n-1}) + \mathbb{E}(x_n \mid \tilde{z}_n) \tag{6.92}$$

The first term is just the prediction of x_n based on $z_{0:n-1}$ to be denoted as

$$\begin{aligned}
\hat{x}_{n|n-1} &= \mathbb{E}(x_n \mid \mathbf{Z}_{n-1}) \\
&= \mathbb{E}(ax_{n-1} + u_n \mid \mathbf{Z}_{n-1}) \\
&= a\mathbb{E}(x_{n-1} \mid \mathbf{Z}_{n-1}) \\
&= a\hat{x}_{n-1|n-1}
\end{aligned} \tag{6.93}$$

where we use the property $\mathbb{E}(u_n) = 0$ since u_n is independent of all

$$w_n, x_{0:n-1}, z_{0:n-1}, u_{0:n-1}.$$

To determine $\mathbb{E}(x_n \mid \tilde{z}_n)$, it is actually the estimator of x_n conditional on $\tilde{x}_n$. In linear form, we have

$$\mathbb{E}(x_n \mid \tilde{z}_n) = \kappa_n \tilde{z}_n \tag{6.94}$$

$$= \kappa_n(z_n - \hat{z}_{n|n-1}) \tag{6.95}$$

where

$$\kappa_n = \frac{\mathbb{E}(x_n \tilde{z}_n)}{\mathbb{E}(\tilde{z}_n^2)} \tag{6.96}$$

via the MMSE estimator for jointly Gaussian θ and y

$$\hat{\theta} = C_{\theta y} C_{yy}^{-1} y = \frac{\mathbb{E}(\theta y)}{\mathbb{E}(y^2)} y$$

Recalling the observation equation $z_n = x_n + w_n$ and property (ii) in the Lemma gives

$$\hat{z}_{n|n-1} = x_{n|n-1} + \hat{w}_{n|n-1} = x_{n|n-1} \qquad (6.97)$$

as w_n is independent of $z_{0:n-1}$ and zero mean. Therefore, (6.95) becomes

$$\mathbb{E}(x_n \mid \tilde{z}_n) = \kappa_n(z_n - \hat{x}_{n|n-1}) \qquad (6.98)$$

Together with (6.92), we have

$$\hat{x}_{n|n} = \hat{x}_{n|n-1} + \kappa_n(z_n - \hat{x}_{n|n-1}) \qquad (6.99)$$

Since $\mathbb{E}\left[w_n(x_n - \hat{x}_{n|n-1})\right] = 0$ due to w_n being uncorrelated with $x_{0:n-1}$ and past observations,

$$\tilde{z}_n = z_n - \hat{z}_{n|n-1} = z_n - \hat{x}_{n|n-1} \qquad (6.100)$$

Since the innovation is uncorrelated with past observations and thus the prediction $\hat{x}_{n|n-1}$ formed by the past observations,

$$\mathbb{E}\left[x_n(z_n - \hat{x}_{n|n-1})\right] = \mathbb{E}\left[(x_n - \hat{x}_{n|n-1})(z_n - \hat{x}_{n|n-1})\right] \qquad (6.101)$$

According to (6.96) and above two equations,

$$\kappa_n = \frac{\mathbb{E}\left[x_n(z_n - \hat{x}_{n|n-1})\right]}{\mathbb{E}\left[(z_n - \hat{x}_{n|n-1})^2\right]} \qquad (6.102)$$

$$= \frac{\mathbb{E}\left[(x_n - \hat{x}_{n|n-1})(z_n - \hat{x}_{n|n-1})\right]}{\mathbb{E}\left[(x_n - \hat{x}_{n|n-1} + w_n)^2\right]} \qquad (6.103)$$

$$= \frac{\mathbb{E}\left[(x_n - \hat{x}_{n|n-1})^2\right]}{\sigma_n^2 + \mathbb{E}\left[(x_n - \hat{x}_{n|n-1})^2\right]} \qquad (6.104)$$

Please note the numerator of (6.104) is just the MSE of one-step prediction, which can be defined as

$$M_{n|n-1} = \mathbb{E}\left[(x_n - \hat{x}_{n|n-1})^2\right]$$
$$= \mathbb{E}\left[(ax_{n-1} + u_n - \hat{x}_{n|n-1})^2\right] \qquad (6.105)$$
$$= \mathbb{E}\left[(a(x_{n-1} - \hat{x}_{n-1|n-1}) + u_n)^2\right]$$

where the system equation and (6.93) are used in above derivations. Since u_n is uncorrelated with x_{n-1},

$$\mathbb{E}\left[(x_{n-1} - \hat{x}_{n-1|n-1})u_n\right] = 0$$

We obtain

$$M_{n|n-1} = a^2 M_{n-1|n-1} + \sigma_u^2 \qquad (6.106)$$

toward the recursive relationship

$$
\begin{aligned}
M_{n|n} &= \mathbb{E}\left[(x_n - \hat{x}_{n|n})^2\right] \\
&= \mathbb{E}\left[(x_n - \hat{x}_{n|n-1} - \kappa_n(z_n - \hat{x}_{n|n-1}))^2\right] \\
&= \mathbb{E}\left[(x_n - \hat{x}_{n|n-1})^2\right] - 2\kappa_n \mathbb{E}\left[(x_n - \hat{x}_{n|n-1})(z_n - \hat{x}_{n|n-1})\right] \\
&\quad + \kappa_n^2 \mathbb{E}\left[(z_n - \hat{x}_{n|n-1})^2\right] \\
&= M_{n|n-1} - 2\kappa_n^2(M_{n|n-1} + \sigma_n^2) + \kappa_n^2 \frac{M_{n|n-1}}{\kappa_n} \\
&= (1 - \kappa_n) M_{n|n-1}
\end{aligned}
$$

$$(6.107)$$

where the second equality comes from (6.99); the fourth equality uses the definition of κ_n; the fact that (6.104) implies

$$\kappa_n = \frac{M_{n|n-1}}{M_{n|n-1} + \sigma_n^2} \qquad (6.108)$$

After lengthy and tedious derivations, the scalar Kalman filter turns out to be quite simple and intuitive. The key factors of *scalar Kalman filter* are summarized in the following Proposition.

Proposition 5. *(Scalar Kalman Filter)* **Prediction:**

$$\hat{x}_{n|n-1} = a\hat{x}_{n-1|n-1}$$

MMSE of Prediction:

$$M_{n|n-1} = a^2 M_{n-1|n-1} + \sigma_u^2$$

Kalman Gain:

$$\kappa_n = \frac{M_{n|n-1}}{\sigma_n^2 + M_{n|n-1}}$$

Correction:

$$\hat{x}_{n|n} = \hat{x}_{n|n-1} + \kappa_n(z_n - \hat{x}_{n|n-1})$$

MMSE:

$$M_{n|n} = (1 - \kappa_n) M_{n|n-1}$$

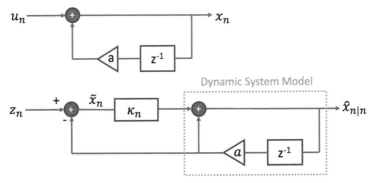

Figure 6.6   Scalar Kalman Filter, in which the dynamic system model is shown in the upper half and also embedded as a part of the Kalman filter.

Figure 6.6 depicts the realization of the scalar Kalman filter, in which the Gauss-Markov signal model (6.81) is shown in the upper and serves as the model of the target dynamic system. Please note the roles of innovation and filter gain. Such a simple realization is easy to implement in hardware or software for wide-range applications without any *a priori* knowledge of system statistics.

6.4.3 Extended Kalman Filter

To relieve the assumptions of linear state transitions and linear measurements embedded in Gaussian noise, the *extended Kalman filter* (EKF) proceeds on the assumptions of the next state probability and the measurement probabilities are governed by nonlinear functions as

$$x_t = g(u_t, x_{t-1}) + \epsilon_t \tag{6.109}$$

$$z_t = h(x_t) + n_t \tag{6.110}$$

The fact that Gaussian input to a linear system generates jointly Gaussian warrantees Gaussian beliefs in Kalman filters. Above nonlinear equations implies that it is not possible to obtain closed-form Bayes filter. The basic concept behind EKF is to compute an approximation to the true belief and to represent such approximation by a Gaussian, particularly for $\mathbb{B}(x_t)$ with mean μ_t and covariance Σ_t. A straightforward way to facilitate such approximation is *linearization*.

Linearization approximates g by a linear function that is tangent to g at the mean of the Gaussian. By projecting the Gaussian through this linear approximation, the posterior is therefore Gaussian. As a matter of fact, as g is linearized, the mechanism of belief propagation would be equivalent to that of a Kalman filter. Similar principle can be applied to h. From calculus, *Taylor expansion* supplies a method of approximation and thus linearization for general nonlinear continuous functions, and EKF can utilize such a method. By Taylor expansion, near the mean of Gaussian belief,

$$g(u_t, x_{t-1}) \approx g(u_t, \mu_{t-1}) + g'(u_t, \mu_{t-1})(x_{t-1} - \mu_{t-1}) \qquad (6.111)$$
$$= g(u_t, \mu_{t-1}) + \qquad G_t(x_{t-1} - \mu_{t-1}) \qquad (6.112)$$

where

$$G_t = g'(u_t, x_{t-1}) = \frac{\partial g(u_t, x_{t-1})}{\partial x_{t-1}} \qquad (6.113)$$

Assuming to be Gaussian, the next state probability is approximated as

$$p(x_t \mid u_t, x_{t-1})$$
$$\approx |\, 2\pi R_t \,|^{-1/2}$$
$$\times e^{-\frac{1}{2}[x_t - g(u_t,\mu_{t-1}) - G_t(x_{t-1}-\mu_{t-1})]^T R_t^{-1}[x_t - g(u_t,\mu_{t-1}) - G_t(x_{t-1}-\mu_{t-1})]}$$
$$(6.114)$$

G_t is often called the *Jacobian*. An EKF implements the exact same linearization for the measurement function h. The Taylor expansion is developed around $\bar{\mu}_t$, the state deemed most likely by the robot at the time it linearizes h:

$$h(x_t) \approx h(\bar{\mu}_t) + h'(\bar{\mu}_t)(x_t - \bar{\mu}_t) = h(\bar{\mu}_t) + H_t(x_t - \bar{\mu}_t) \qquad (6.115)$$

We have

$$p(z_t \mid x_t) = |\, 2\pi R_t \,|^{-1/2} e^{-1/2}$$
$$\times e^{-\frac{1}{2}[z_t - h(\bar{\mu}_t) - H_t(x_t - \bar{\mu}_t)]^T Q_t^{-1}[z_t - h(\bar{\mu}_t) - H_t(x_t - \bar{\mu}_t)]}$$
$$(6.116)$$

Proposition 6. Extended Kalman Filter Algorithm: *Input:* $\mu_{t-1}, \Sigma_{t-1}, u_t, z_t.$

$$\bar{\mu}_t = g(u_t, \mu_{t-1}) \qquad (6.117)$$
$$\bar{\Sigma}_t = G_t \Sigma_{t-1} G_t^T + R_t \qquad (6.118)$$
$$K_t = \bar{\Sigma}_t H_t^T (H_t \bar{\Sigma}_t H_t^T + Q_t)^{-1} \qquad (6.119)$$
$$\mu_t = \bar{\mu}_t + K_t(z_t - h\bar{\mu}_t) \qquad (6.120)$$
$$\Sigma_t = (I - K_t H_t)\bar{\Sigma}_t \qquad (6.121)$$

Return: $\mu_t, \Sigma_t.$

Remark: Kalman filter is an optimal linear estimation. However, to deal with nonlinearity that is common in engineering, extended Kalman filters can be therefore applied at the price of losing optimality and sensitive to initial state selection.

Further Reading: For more understanding about random processes and basic statistical signal processing, [1] is a straightforward textbook worth reading. Complete study of principles and techniques for estimation can be found in a classic book [2]. [4] has detailed contents about recursive estimation and readers of particular interest in robotics are encouraged to read this book in detail.

References

[1] R.M. Gray, L.D. Davisson, *An Introduction to Statistical Signal Processing*, Cambridge University Press, 2004

[2] S.M. Kay, *Fundamentals of Statistical Signal Processing, Vol. I: Estimation*, Prentice Hall, 1993.

[3] A.R. Ribeiro, G.B. Giannakis, "Bandwidth-Constrainted Distributed Estimation For Wireless Sensor Networks - Part I: Gaussian Case", *IEEE Tr, on Signal Processing*, vol. 54, no. 3, pp. 1131-1143, March 2006.

[4] S. Thrun, W. Burgard, D. Fox, *Probabilistic Robotics*, MIT Press, 2006.

[5] L.R. Rabiner, B.H. Juang, "An Introduction to Hidden Markov Models", *IEEE ASSP Magazine*, Jan. 1986.

7

Localization

Autonomous mobile robots (AMRs) can generally take actions to move and *localization* is the general problem to determine the pose of a robot relative to a given environment, where a given map for the environment may or may not be available. Localization is also known as *position estimation* or *position tracking* and is a fundamental perception technology since almost all robotic tasks require the knowledge of the location of the robots. There are two common types of localization problems: *mobile robot localization* and *sensor network localization*.

Since a robot's private (or local) reference coordinate shall be aligned to public (or global) reference coordinate, localization can be generally viewed as a coordinate transformation. When the global map (or public reference) is available to describe the global coordination system, localization becomes the process to establish the correspondence between the map (i.e. public reference) and the robot's local coordinate system (i.e. private reference). The knowledge of such coordinate transformation enables the robot to interpret location of interest within its own (i.e. local) coordinate system, which is a prerequisite for robot navigation. Knowing the pose of the robot is sufficient to determine the coordinate transform of a robot with fixed position. Nevertheless, the pose is usually not directly known and is typically inferred from the sensor data, while a single sensor measurement is usually insufficient to determine the pose.

Consequently, there are three general scenarios of localization:

Position Tracking When the initial pose of the robot is known, localization of the robot can be achieved by assuming the small noise (or uncertainty) of robot motion. Such uncertainty of pose is often approximated by a unimodal distribution (e.g. Gaussian) and position tracking is then a local problem.

189

Global Localization When the initial pose is unknown, the robot is initially placed in the environment, lacking the knowledge of location. Bounded pose error and unimodal probability distribution are therefore inappropriate assumptions, to make this class of localization difficulty.

Kidnapped Robot Due to possible system error, the kidnapped robot might believe it knows the location but actually not. Compared with global localization that robot at least knows that it does not know the location, kidnapped robot problem is even more challenging, which is related to the capability of recovering from failures as the essential feature of autonomous robots.

The robots can operate in *static environments* or *dynamic environments*. The *active localization* algorithm can control the movement of a robot. As a contrast, *passive localization* just observes the operation of a robot.

7.1 Localization By Sensor Network

It is straightforward to use sensors that forms a sensor network to accomplish localization of a robot. Generally speaking, highly accurate localization consists of two phases:

(a) accurate ranging (i.e. estimating the distance between the robot and a sensor)
(b) exact localization (i.e. determining the exact location of an unknown robot, with respect to the anchor's known position)

It is interesting to note that the application of estimation theory or statistical signal processing can assist the facilitation of artificial intelligence in robotics.

7.1.1 Time-of-Arrival Techniques

Since the distance information is embedded in the propagation delay between the transmitter and receiver, which is ideally the product of propagation delay and speed of light, *time-of-arrival* (TOA) serves as the most fundamental technique for localization. TOA works if the receiver acts as an anchor (node) that the transmitter aligns, or if the receiver is perfectly synchronized to the anchor (node). Figure 7.1 illustrates the fundamental TOA estimation system, where the green square represents the true location of the robot and the red circle represents the estimator of its location.

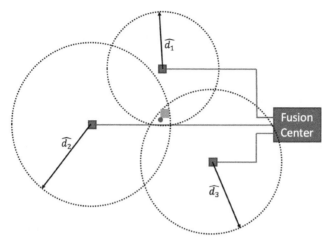

Figure 7.1 System Model of Localization Based on TOA Technique (green square: robot; red circle: estimator of robot's location).

Suppose there are N location sensors with a fusion center to estimate the location of a robot, $\hat{\mathbf{x}} = (\hat{x}, \hat{y})^T$, where $\hat{\mathbf{x}}_{\mathbf{i}} = (\hat{x}_i, \hat{y}_i)^T, i = 1, \cdots, N$ denotes the known position of the ith sensor, and $\hat{d}_i$ is the measured distance between the robot and the ith sensor. Such measured distance can be modeled as

$$\hat{d}_i = d_i + b_i + n_i = c\tau_i, i = 1, \cdots, N \qquad (7.1)$$

where τ_i denotes TOA of the signal at the ith sensor; c denotes the speed of light (i.e. radio propagation in free space); d_i denotes the distance between the robot and the ith sensor; $n_i \sim G(0, \sigma_i^2)$ is the additive white Gaussian noise from the measurement; b_i represents the bias introduced due to the blockage of the direct path and

$$b_i = \begin{cases} 0, & if\ the\ ith\ sensor\ in\ line\ of\ sight\ (LOS) \\ \psi_i, & if\ the\ ith\ sensor\ in\ non-line\ of\ signt\ (NLOS) \end{cases} \qquad (7.2)$$

Then, define

$$\mathbf{d} = \mathbf{d}(\mathbf{x}) = (d_1, d_2, \cdots, d_n)^T \qquad (7.3)$$

$$\hat{\mathbf{d}} = (\hat{d}_1, \cdots, \hat{d}_N \qquad (7.4)$$

$$\mathbf{b} = (b_1, \cdots, b_N) \qquad (7.5)$$

$$\mathbf{Q} = \mathbb{E}\left[\mathbf{n}\mathbf{n}^T\right] = \text{diag}\left[\sigma_1^2, \cdots, \sigma_N^2\right]^T \qquad (7.6)$$

For ideal measurement without noise and bias, the true distance d_i between the robot and the ith sensor defines a circle around the ith sensor.

$$(x - x_i)^2 + (y - y_i)^2 = d_i^2, i = 1, \cdots, N \qquad (7.7)$$

The possible location of the robot lies inside the intersection of these circles as Figure 7.1. However, the noisy measurements and the NLOS bias yield another inconsistent equation

$$(x - x_i)^2 + (y - y_i)^2 = \hat{d}_i^2, i = 1, \cdots, N \qquad (7.8)$$

An effective estimator of robot's location is therefore wanted. From earlier chapter regarding the estimation, this can be precisely treated as the *maximum likelihood estimation* (MLE). By ignoring the NLOS issue, we assume $b_i = 0, \forall i$ and independent measurements among these N sensors. The likelihood function is

$$p(\hat{\mathbf{d}} \mid \mathbf{x}) = \prod_{i=1}^{N} \frac{1}{\sqrt{2\pi\sigma_i^2}} e^{\frac{(\hat{d}_i - d_i)^2}{2\sigma_i^2}} \qquad (7.9)$$

$$= (2\pi)^{-N/2} \left[\det(\mathbf{Q})\right]^{1/2} e^{-\frac{1}{2}[\hat{\mathbf{d}} - \mathbf{d}(\mathbf{x})]^T \mathbf{Q}^{-1}[\hat{\mathbf{d}} - \mathbf{d}(\mathbf{x})]} \qquad (7.10)$$

The MLE of robot's location is derived from

$$\hat{\mathbf{x}}_{ML} = \arg\max_{\mathbf{x}} p(\hat{\mathbf{d}} \mid \mathbf{x}) \qquad (7.11)$$

Generally speaking, implementation of above equation requires a computational intensive search over possible locations. For a special case of $\sigma_i^2 = \sigma^2, \forall i$, the solution of MLE is equivalent to minimizing the log likelihood

$$J \triangleq \left[\hat{\mathbf{d}} - \mathbf{d}(\mathbf{x})\right]^T \mathbf{Q}^{-1} \left[\hat{\mathbf{d}} - \mathbf{d}(\mathbf{x})\right] \qquad (7.12)$$

Again, the necessary condition $\nabla_{\mathbf{x}} J = 0$ yields

$$\sum_{i=1}^{N} \frac{(d_i - \hat{d}_i)(x - x_i)}{d_i} = 0$$

$$\sum_{i=1}^{N} \frac{(d_i - \hat{d}_i)(y - y_i)}{d_i} = 0$$

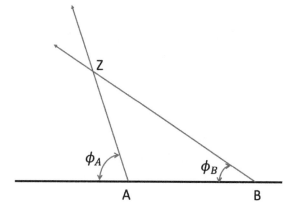

Figure 7.2 Principle of angle of arrival.

which is not possible to obtain **x** in general closed form using a linear least squares (LS) algorithm, and any computationally efficient distributed localization algorithm is still very much wanted.

Remark: In addition to what we are going to introduce later, another class of techniques is pattern matching, in which fingerprint information of the measured radio signals at different geographical locations are utilized. More details can be found in signal processing literature.

7.1.2 Angle-of-Arrival Techniques

The *angle-of-arrival* (AOA) reconstructs the distance between transmitter and receiver based on their angle, which antenna array is typically employed. As shown in Figure 7.2 in 2-D plane, suppose we know the locations of A and B, we can determine the location of Z based on the angles ϕ_A and ϕ_B. AoA is generally sensitive to multi-path fading and thus antenna array and beamforming are usually employed together with AOA technique.

▶ **Exercise:** As Figure 7.2, suppose the coordinates of A and B to be (x_A, y_A) and (x_B, y_B). By measuring the angles ϕ_A and ϕ_B, please determine the 2-D coordinates of point Z (i.e. the location of Z).

Modern implementation of AoA techniques often takes advantage of antenna patterns, thanks to advances in RF and antenna technology. AoA can proceed based on either the amplitude response at the receiver antenna(s) or the phase response at the receiver antenna(s). To utilize the amplitude response of receive antenna for the measurement of AOA, we leverage the

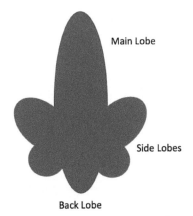

Figure 7.3   Illustration of the horizontal antenna pattern of a typical anisotropic antenna.

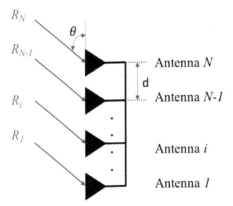

Figure 7.4   Illustration of an antenna array of N elements.

antenna pattern as shown Figure 7.3, particularly the main lobe that implies the strongest antenna gain. By mechanically or electronically rotate (or tilt) the direction of antenna, the direction of radio emission can be identified and thus the angle of signal arrival.

Obviously, the amplitude response of a single receive antenna suffers from noisy observations, signal fluctuations, and accuracy of measurement. Another means is to utilize the phase information of an antenna array as shown in Figure 7.4. This technique is known as *phase interferometry*, deriving AOA measurements from the phase differences in the arrival of a wavefront, while a large receive antenna or an antenna array is typically required.

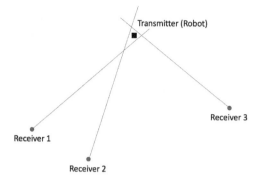

Figure 7.5   Under noisy observations, the bearing lines from three receivers generally can not intersect at the same point.

As Figure 7.4 of N antenna elements, The adjacent antenna elements are separated by a uniform distance d. The distance between a far away transmitter and the ith antenna element is

$$R_i \approx R_1 - (i-1)d\cos\theta \tag{7.13}$$

The signals received by adjacent antenna elements have a phase differ- ence of $2\pi\frac{d\cos\theta}{\lambda}$, which results in the bearing of the transmitter from the measurement of the phase difference. Such a method works well in high signal-to-noise-ratio and further techniques have been developed to improve.

After the measurement, if there is no noise and interference, bearing lines from two or more receivers can intersect to determine a unique location, that is, the location estimate of the transmitter (say, a robot). In the presence of noise, more than two bearing lines may not intersect at a single point. Consequently, statistical algorithms (sometimes called triangulation or fixing methods) are needed to estimate the location of the transmitter.

As Figure 7.5, a 2D AOA localization problem using the bearing mea- surements can be formulated as follows. Let $\mathbf{X}_t = (x_t, y_t)^T$ be the true coordinate vector of the transmitter (i.e. robot) to be estimated by 2D bearing measurements $\mathbf{b} = (b_1, \cdots, b_N)^T$, where N denotes the total number of receivers. Denote $\mathbf{x}_i = (x_i, y_i)^T$ as the coordinates of known location for the ith receiver making bearing measurement. De note the bearings of a transmitter located at $\mathbf{x} = (x, y)^T$ from the known receiver locations as $\phi(\mathbf{x}) = [\phi_1(\mathbf{x}), \cdots, \phi_N(\mathbf{x})]^T$, where $\phi_i(\mathbf{x}), 1 \leq i \leq N$ are related to $\mathbf{x}$ by

$$\tan\phi_i(\mathbf{x}) = \frac{y - y_i}{x - x_i} \tag{7.14}$$

Assume that the measured bearings of the transmitter consist of the true bearings corrupted by additive Gaussian noises $\mathbf{n} = (n_1, \cdots, n_N)^T$ with zero mean and covariance matrix $\mathbf{Q} = diag\{\sigma_1^2, \cdots, \sigma_N^2\}$. That is,

$$\mathbf{b} = \phi(\mathbf{x}_t) + \mathbf{n} \tag{7.15}$$

In a special case that the receivers are identical and much closer to each other than to the transmitter, the variances of bearing measurement errors tend to equal, that is, $\sigma_1^2 = \cdots = \sigma_N^2 = \sigma^2$. The maximum likelihood (ML) estimator of the transmitter's location $\mathbf{x}_t$ is generally given by

$$\mathbf{x}_t = \operatorname{argmin} [\phi(\hat{\mathbf{x}}_t) - \mathbf{b}]^T \mathbf{Q}^{-1} [\phi(\hat{\mathbf{x}}_t) - \mathbf{b}] \tag{7.16}$$

$$= \sum_{i=1}^{N} \frac{[\phi_i(\hat{\mathbf{x}}_t) - b_i]^2}{\sigma_i^2} \tag{7.17}$$

which can be numerically solved by the Newton-Gauss iteration method.

7.1.3 Time-Difference-of-Arrivals Techniques

In the TOA method, there still exists a critical issue to complete, *time delay estimation* (TDE), since we have to use the delay information to compute corresponding distance. This wireless technique has a broad-range of applications in robotics, such as localization, robot-pose, sensing environments, and control of robot actions. In many cases, estimation of distance can be accomplished by either received signal strength (RSS) or estimation of propagation time.

In free space, the strength of received signal can be related to the Friis equation. Suppose the received power $P_r(d)$. Then,

$$P_r(d) = \frac{P_t G_t G_r \lambda^2}{(4\pi)^2 d^2} \tag{7.18}$$

where P_t denotes the transmitted power; G_t and G_r represent the transmit antenna gain and the receive antenna gain, respectively; λ is the wavelength in meters. This free-space model for RSS is rather ideal and thus RSS profiling measurements would be more practical.

As far as TDE, it can be classified into two categories: active and passive. The *active TDE* assumes the signal $s(t)$ to be known, which leads to the following formulation.

$$r(t) = \alpha s(t - D) + w(t), \ 0 \le t \le T \tag{7.19}$$

where $r(t)$ denotes the received waveform with delay D, period of signal T, and additive noise $w(t)$. Its digital signal processing formulation turns to

$$r_n = \alpha s_{n-D} + w_n, \quad n = 0, 1, \cdots, N-1 \qquad (7.20)$$

▶ **Exercise:** To find D in equation (7.20) is equivalent to baseband timing recovery in a digital communication system through an AWGN channel. For a given s_n, $n = 0, 1, \cdots, N-1$, please derive the estimation of D (the most likely value of D among $0, 1, \cdots, N-1$).

For passive TDE (i.e. signal $s(t)$ unknown), we have to rely on multiple receivers, at least two. The signal model to estimate D is therefore,

$$r_1(t) = s(t) + w_1(t) \qquad (7.21)$$
$$r_2(t) = \alpha s(t - D) + w_2(t) \qquad (7.22)$$

where $s(t), w_1(t), w_2(t)$ are stationary, and $s(t)$ us uncorrelated with $w_1(t), w_2(t)$. For *passive TDE*, different from typical digital communication systems, the source (or signal) spectrum is unknown (at most, proximately known). In order to determine D, we compute the cross-correlation

$$R_{r_1, r_2}(\tau) = \mathbb{E}\left[r_2(t)r_2(t - \tau)\right] \qquad (7.23)$$

Assume ergodic processes, we can have

$$\hat{R}_{r_1, r_2}(\tau) = \frac{1}{T - \tau} \int_{\tau}^{T} r_1(t)r_2(t - \tau)dt \qquad (7.24)$$

where T is the observation interval. Above suggests an intuitive cross-correlation TDE as Figure 7.6, which describes the principle of two filtered received waveforms, delay-and-multiply, integrate-square, and peak detection to obtain the TDE.

Obviously, more receivers may indicate more effective estimation. It suggests the *time difference of arrivals* (TDOA) that proceeds based on the difference among TOAs in several receivers (i.e. sensors) to reconstruct the

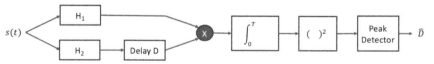

Figure 7.6 Correlation Principle of TDE.

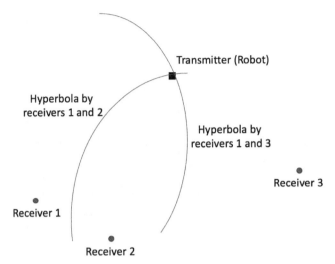

Figure 7.7　Intersecting hyperbolas from three receivers to locate the transmitter (i.e. robot).

transmitter's (i.e. robot's) location. Suppose we have M received sensors, each has the signal model as

$$r_i(t) = \alpha_i s(t - \tau_i) + w_i(t), \ i = 1, \cdots, M \qquad (7.25)$$

where $s(t)$ is the signal of interest; α_i, τ_i are the channel gain (i.e. attenuation) and propagation delay, respectively. Given $r_i(t)$, TDE is to estimate

$$\tau_{i,j} = -\tau_{j,i} = \tau_i - \tau_j, \ i > j, i, j = 1, \cdots, M \qquad (7.26)$$

Although there are $M(M - 1)/2$ delays, there are only $M - 1$ non-redundant parameters due to the fact $\tau_{i,j} = \tau_{i,k} - \tau_{j,k}, k \neq i, j$. An example of the redundant set is $\{\tau_{i,1}\}, i = 2, \cdots, M$. This form the principle of *time difference of arrival* (TDOA), to expect more precise estimation as shown in Figure 7.7.

7.2 Mobile Robot Localization

Since localization usually involves uncertainty, probabilistic approach is again useful. We start from probabilistic localization algorithms that are variants of the Bayes filter in Chapter 6, which is called *Markov localization*. Markov location algorithm requires a map $\mathcal{M}$ as input of the algorithm. With the input variables $\mathbb{B}(x_{t-1}), u_t, z_t, \mathcal{M}, \forall x_t$, the algorithm computes

$$\bar{\mathbb{B}}(x_t) = \int p(x_t \mid u_t, x_{t-1}, \mathcal{M}) \mathbb{B}(x_{t-1}) dx \qquad (7.27)$$

$$\mathbb{B}(x_t) = c_{ml} p(z_t \mid x_t, \mathcal{M}) \bar{\mathbb{B}}(x_t) \qquad (7.28)$$

As the Bayes filter in Section 6.3, Markov localization transforms a probabilistic belief at time $t-1$ into a belief at time t. Markov localization can be applied to the global localization problem, the position tracking problem, and the kidnapped robot problem in static environments.

The initial belief, $\mathbb{B}(x_0)$, indicates the initial knowledge of the robot pose/location. Its setting depends on the type of localization problem.

Position Tracking If the initial pose of robot is known as $\bar{x}_0$, $\mathbb{B}(x_0)$ is initialized to have all probability mass at this point.

$$\mathbb{B}(x_0) = \begin{cases} 1, & x_0 = \bar{x}_0 \\ 0, & \text{otherwise} \end{cases} \qquad (7.29)$$

In engineering practice, the initial pose is known as an approximation, and thus the belief is usually described as a narrow-centered Gaussian distribution around $\bar{x}_0$. Referring (6.71),

$$\mathbb{B}(x_0) = \det(2\pi\Sigma)^{-\frac{1}{2}} \exp\left[-\frac{1}{2}(x_0 - \bar{x}_0)^T \Sigma^{-1}(x_0 - \bar{x}_0)\right] \qquad (7.30)$$

Global Localization If the initial pose/location is unknown, $\mathbb{B}(x_0)$ is initialized by a uniform distribution over all legislative poses or all locations in the map (i.e. reference system) as a least favorable distribution.

$$\mathbb{B}(x_0) = \frac{1}{|X|} \qquad (7.31)$$

where $|X|$ denotes the cardinality or measure of all possible poses/locations.

Partial knowledge of the robot?s pose/location can easily be transformed into an appropriate initial probability distribution, possibly with the aid of inference.

7.3 Simultaneous Localization and Mapping

After introducing mobile robot localization, when the mobile robot has no prior information about the environment nor its pose, *simultaneous localization and mapping* (SLAM) arises as an important technology in robotics. SLAM is a process by which a mobile robot can establish own private reference system (i.e. map) corresponding to the environment, and deduce its location using this private reference system (i.e. map) at time same time, to align with the public reference system (or map).

In SLAM, the robot develops a map of its environment while localizing itself relative to the map. From the statistical perspective, there are two primary forms of the SLAM:

(a) Online SLAM: Online SLAM is to estimate the posterior over the momentary pose along the map, $p(x_t, \mu \mid z_{1:t}, u_{1:t})$, where x_t is the pose at time t, $\mathcal{M}$ is the map, and $z_{1:t}$ and $u_{1:t}$ are the measurements and controls up to time t, respectively.

(b) Full SLAM: Full SLAM is to compute a posterior over the entire path $x_{1:t}$ along the map, $p(x_{1:t}, \mathcal{M} \mid z_{1:t}, u_{1:t})$, instead of just the current pose x_t.

The online SLAM results from integrating over past poses from the full SLAM:

$$p(x_t, \mathcal{M} \mid z_{1:t}, u_{1:t}) = \int \cdots \int p(x_{1:t}, \mathcal{M} \mid z_{1:t}, u_{1:t}) dx_1 \cdots dx_{t-1} \quad (7.32)$$

Shown in Figure 7.8, a typical SLAM problem contains a continuous problem and a discrete component. The continuous estimation problem pertains to the location of the objects in the map and the robot's own pose variables. Objects are often known as landmarks in feature-based representation, or they might be object patches detected by range sensors. The discrete nature has to do with correspondence: When an object is detected, a SLAM algorithm must reason about the relation of this object to previously detected objects, while such reasoning is typically discrete (i.e. true or not).

7.3.1 Probabilistic SLAM

To practically develop the concept of SLAM, we have to take advantage of landmarks in a reference system (or a map). The trajectory of a mobile robot and locations of all landmarks can be online estimated without the need of any *a priori* knowledge. Consider a mobile robot moving through the environment

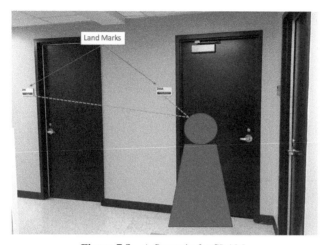

Figure 7.8 A Scenario for SLAM.

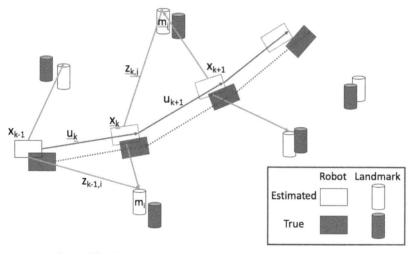

Figure 7.9 True and estimated locations of robots and landmarks.

and taking observations of a number of (unknown) landmarks using its on-board sensors as Figure 7.9. Please note that the locations of landmarks might be known in the map $\mathcal{M}$, or unknown. Let us re-iterate the notations:

- x_t: the state (vector) for location and orientation/pose of the mobile robot
- u_t: the control (vector), applied at time $t - 1$ to drive the robot to state x_t at time t,

- m_i: the location coordinate (vector) of the ith landmark with time-invariant true location
- z_{it}: an observation (vector) by the robot about the location of the ith landmark at time t. For multiple landmark observations at time t, the observations are denoted as z_t.
- $m = \{m_1, \cdots, m_M\}$ denotes the set of all landmarks.

The probabilistic formulation of SLAM problem is to compute

$$p(x_t, m \mid z_{0:t}, u_{0:t}, x_0), \ \forall t$$

This probability distribution represents the joint *posteriori* density of the landmark locations and robot state at time t, given the recored observations and control inputs up to time t with initial state of the robot.

Generally speaking, a recursive solution suitable for online realization is desirable. The estimate $p(x_{t-1}, m \mid z_{0:t-1}, u_{0:t-1})$ at time $t-1$, the joint posterior, following the control u_t and new observation z_t, can be used to compute by Bayes theorem, which requires a state transition model and an observation model.

The *observation model* represents the probability of obtaining an observation z_t when the locations of the robot and landmarks are known, and can be described as

$$p(z_t \mid x_t, m)$$

It is reasonable to assume that observations are conditionally independent given the map and current state of the robot, once robot's location and map are defined.

The *motion model* for a robot can be developed in terms of probability distribution on state transitions that are assumed to be a Markov process as

$$p(x_t \mid x_{t-1}, u_t)$$

in which the current state x_t depends only on the immediately preceding state x_{t-1} and the control u_t, independent of the observations and the map.

Proposition (SLAM Algorithm): The SLAM algorithm can be implemented as a two-step recursive prediction (time-update) and correction (measurement-update) form:

Time-Update

$$p(x_t, m \mid z_{0:t-1}, u_{0:t}, x_0)$$
$$= \int p(x_t \mid x_{t-1}, u_t) \cdot p(x_{t-1}, m \mid z_{0:t-1}, u_{0:t-1}, x_0) dx_{t-1} \qquad (7.33)$$

Measurement-Update

$$p(x_t, m \mid z_{0:t}, u_{0:t}, x_0) = \frac{p(z_t \mid x_t, m) p(x_t, m \mid z_{0:t-1}, u_{0:t}, x_0)}{p(z_t \mid z_{0:t-1}, u_{0:t})} \quad (7.34)$$

Remark: A map $\mathcal{M}_R$ can be constructed by fusing observations (of landmarks) from different locations, which is obviously a private reference/map for robot itself, which is not necessarily identical to the global reference (or true map), $\mathcal{M}$, but hopefully to be finally consistent.

Remark: Referring Figure 7.9, the error between estimated and true locations for a landmark is primarily due to the knowledge of robot's location, which suggests the errors of the estimates in landmark location are highly correlated. Or, the relative locations between two landmarks may be practically known with high accuracy. Consequently, an insight of SLAM is to realize the correlations between the estimates of the landmark. In probability, $p(m)$ is monotonically more converging. As Figure 7.9, the robot at the state x_k observes two landmarks m_i and m_j. The relative location of observed landmarks is clearly independent of the private coordinate system (or map) of the robot, and successive observations from this fixed location would yield further independent measurements of the relative relationship between landmarks. When the robot moves to next location x_{k+1} and observes landmark m_j, this allows the estimated location of the robot and landmark being updated relative to the previous location x_k. In turn, this propagates back to update landmark m_i, even though this landmark might not be seen from the new location, because the two landmarks are highly correlated (their relative location is well known) from previous measurements. Furthermore, the fact that the same measurement data is used to update these two landmarks makes them more correlated. This convergence concludes that the observations made by the robot can be considered as nearly independent measurements of the relative location between landmarks

7.3.2 SLAM with Extended Kalman Filter

A common solution to the probabilistic SLAM involves finding appropriate representations for the observation model and the motion model, to efficiently compute the prior and posterior distribution in the time-update and measurement-update. In terms of state-space model with additive Gaussian noise, *extended Kalman filter* (EKF) widely serves the purpose of implementing SLAM, known as *EKF-SLAM*.

In the motion model $p(x_t \mid x_{t-1}, u_t)$,

$$x_t = f(x_{t-1}, u_t) + w_t \tag{7.35}$$

where $f(\cdot)$ represents robot's kinetics and w_t is additive, zero mean, independent Gaussian disturbance with covariance Q_t.

Regarding the observation model $p(z_t \mid x_t, m)$,

$$z_t = h(x_t, m) + v_t \tag{7.36}$$

where $h(\cdot)$ represetns the geometry of observation and v_t is additive, zero mean, independent Gaussian error with covariance R_t.

The standard EKF method can be applied to compute the mean

$$\begin{bmatrix} \hat{x}_{t|t} \\ \hat{m}_t \end{bmatrix} = \mathbb{E}\begin{bmatrix} x_t \\ m \end{bmatrix} \mid z_{0:t} \end{bmatrix} \tag{7.37}$$

and covariance

$$\Psi_{t|t} = \begin{bmatrix} \Psi_{xx} & \Psi_{xm} \\ \Psi_{xm} & \Psi_{mm} \end{bmatrix} = \mathbb{E}\left[\begin{pmatrix} x_t - \hat{x}_t \\ m - \hat{m}_t \end{pmatrix} \begin{pmatrix} x_t - \hat{x}_t \\ m - \hat{m}_t \end{pmatrix}^T \mid z_{0:t} \right] \tag{7.38}$$

of the joint posterior distribution $p(x_t, m \mid z_{0:t}, u_{0:t}, x_0)$.

Proposition (EKF-SLAM Algorithm):

Time-Update

$$\hat{x}_{t|t-1} = f(\hat{x}_{t-1|t-1}, u_t) \tag{7.39}$$

$$\Psi_{xx,t|t-1} = \nabla f \Psi_{xx,t-1|t-1} \nabla f^T + Q_t \tag{7.40}$$

where ∇f is the Jacobian of f evaluated at the estimate $\hat{x}_{t-1|t-1}$. There is hardly a need to perform time-update for stationary landmarks.

Observation-Update

$$\begin{bmatrix} \hat{x}_{t|t} \\ \hat{m}_t \end{bmatrix} = \begin{bmatrix} \hat{x}_{t|t-1} \hat{m}_{t-1} \end{bmatrix} + \Upsilon_t \begin{bmatrix} z_t - h(\hat{x}_{t|t-1}, \hat{m}_{t-1}) \end{bmatrix} \tag{7.41}$$

$$\Psi_{t|t} = \Psi_{t|t-1} - \Upsilon_t \Xi_t \Upsilon_t^T \tag{7.42}$$

where

$$\Xi_t = \nabla h \Psi_{t|t-1} \nabla h^T + R_t \tag{7.43}$$

$$\Upsilon_t = \Psi_{t|t-1} \nabla h^T \Xi_t^{-1} \tag{7.44}$$

and ∇h is the Jacobian of h evaluated at $\hat{x}_{t|t-1}$ and $\hat{m}_{t-1}$.

Remark: The convergence of the map suggests $\Psi_{mm,t}$ toward zero. The computational complexity has been widely explored in literature such as [7].

Remark: Generally speaking, the EKF SLAM algorithm applies the EKF to online SLAM using maximum likelihood data association, subject to approximations and assumptions:

- Feature-based map in the EKF is composed of point landmarks. Consequently, EKF SLAM requires significant feature detections, and sometimes uses artificial beacons or landmarks as features.
- As any EKF algorithm, EKF SLAM makes a Gaussian noise assumption for the robot motion and the perception.
- The EKF SLAM algorithm, just like the EKF localizer, can only process positive sightings of landmarks. It cannot process negative information that arises from the absence of landmarks in a sensor measurements.

7.3.3 SLAM Assisted by Stereo Camera

One of the immediate realizations of Figure 7.9 is to take advantage of visionary methodology, by the geometric analysis of images or videos. Recall the fact that predators tend to move using the visual information from two eyes in front, which suggests the application of a *stereo camera* to refine its depth awareness as well as geometric localization in any situation where the camera moves. An inertial measurement unit combines a variety of sensors with gyroscopes to detect both rotation and movement in 3 axes, as well as pitch, yaw, and roll. The output of the camera used in the robot implementation includes:

- frames of visual information
- position of the camera
- orientation of the camera
- linear velocity of the camera
- depth of the landmark

Figure 7.10 depicts the SLAM technique utilizing a stereo camera. We use an example to illustrate this technique. Without loss of generality, suppose we intend to design a lawn mower robot working in the environment as Figure 7.11 with house and trees. The target working area is marked by four landmarks with boundaries to be detected that is the primary task for localization.

The landmarks as well as the obstacles can be detected by image processing. The purpose of detecting landmarks is to determine the boundaries of

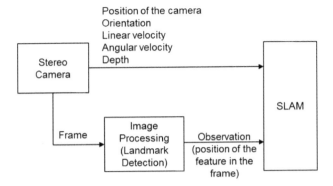

Figure 7.10 Using stereo camera for SLAM.

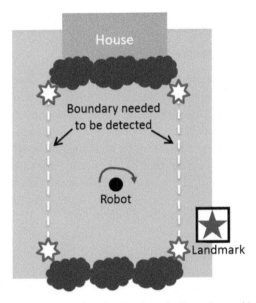

Figure 7.11 Lawn with four landmarks to indicate the working area.

working area, and the purpose of detecting obstacles is to avoid unexpected executions damaging the flowers and trees, without hitting the stones or blocks damaging the robot. As the description of Chapter 9, the edges in an image can supply a lot of information to recognize an object. When the camera turns on, consecutive frames are captured with a predefined frame rate (i.e. frames per second). The output of the image processing is a control signal that informs the next functionality *data fusion* to fuse sensor data to localize

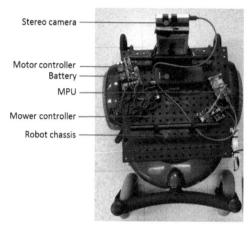

Figure 7.12   A lawn mower robot using stereo camera and SLAM, and RL to control robot's movement, where MPU means micro-processor unit.

the landmark, which will be discussed in later chapter about multimodal data fusion. Once the feature is detected, the position of the detected landmark will be passed to SLAM.

For a cleaning robot working indoor, SLAM is rather straightforward as the boundary detection is rather intuitive by hitting an obstacle like a wall. For a mobile robot operating outdoor, localization is more complicated. A straightforward way is to use GPS, which is not precise enough in many cases. Differential GPS technique together with street map could serve the purpose of car navigation. However, for general applications of mobile robots on campus or factory, including lawn mowing, alternative technique such as SLAM using landmarks would be required for localization. Figure 7.12 depicts a lawn mower robot developed by students at the Department of Electrical Engineering, University of South Florida.

To implement SLAM, a large state vector stacks the states of the robot and M landmarks, given by $\mathbf{x}_t = (m_{R,t}, m_{1,t}, \cdots, m_{M,t})^T$. $m_{R,t} = (x_{R,t}, y_{R,t}, \theta_{R,t}, v_{R,t}, \phi_{R,t})$ denotes the state of camera mounted on top of the mobile robot at time t, where $(x_{R,t}, y_{R,t})$ is the 2-D position by assuming plat terrain; $\theta_{R,t}$ represents the orientation; $(v_{R,t}, \phi_{R,t})$ indicates the linear and angular velocities. $m_{m,t} = (x_{m,t}, y_{m,t}, d_{m,t})$, $m = 1, \cdots, M$ denotes the state vector of the mth landmark at time t, while $(x_{m,t}, y_{m,t})$ and $d_{m,t}$ represent the 2-D position and the depth of the landmark. According to Figure 7.11, there are $M = 4$ landmarks, while each uses a human face as the feature of the landmark in implementation.

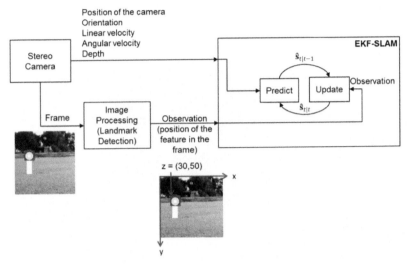

Figure 7.13   Implementation of EKF-SLAM using the state of the stereo camera and the position of the detected landmark in the frame.

As shown in Figure 7.13, EKF-SLAM is employed to implement the prediction and update in a recursive manner. In the prediction stage, EKF-SLAM uses the camera's linear and angular velocities to predict the next state $\hat{x}_{t|t-1}$. In the update stage, the measurement residual $\mathbf{y}_t = \mathbf{z}_t - f(\mathbf{x}_{t|t-1})$ is computed to update the state $\mathbf{x}_{t|t} = \mathbf{x}_{t|t-1} - \kappa_t \mathbf{y}_t$, where $f(\cdot)$ is the nonlinear observation function; κ_t is the Kalman gain; and $\mathbf{z}_t = (x_t, y_t)$ denotes the observation.

7.4 Network Localization and Navigation

The major purposes of localization aims at smooth navigation of mobile robots. Modern robotics requires real-time processing and high precision. Cooperation among nodes (sensors or robots) at physical layer improves accuracy and reliability in localization and consequent navigation of robots. The localization and navigation process of a robot typically consists of two phases such that actuator can conduct the actions of navigation:

(a) measurement phase, in which robots/agents make intra- and inter-node measurements using different sensors

(b) location update phase, in which robots/agents infer their own positions using an algorithm that incorporates both prior knowledge of their positions and new measurements.

The accuracy of localization is typically measured by the mean squared error (MSE) of the position estimate, which means the squared Euclidean distance between the estimated position $\hat{x}$ and true position x, as $e^2(x) = \|\hat{x} - x\|^2$. A global performance metric evaluated over the entire localization area and time is the *localization error outage* (LEO) defined by

$$P_{out} = P\{e^2(x) > e_0^2\} \tag{7.45}$$

where e_0 denotes the maximum allowable position estimation error and the probability is evaluated over the ensemble of all possible spatial area and time duration. For the navigation of a mobile robot, localization update rate (i.e. the number of position estimations per second) is another important system parameter.

Earlier in this book, we know that multiple sensors cooperate to generate estimate of a robot's location. The concept of cooperation has been applied to wireless sensor networks (WSNs), where distributed sensors work together to reach a consensus about the environment or to estimate a spatiotemporal process based on their local measurements. Consider a network with anchors and N_a agents who are equipped with multiple sensors to provide intra- and inter-node measurements for localization and navigation. Using these intra- and inter-node measurements, denoted as $z = [z_{self} z_{rel}]$, the agents infer their locations $x = [x_1, \cdots, x_{N_a}]$, while the accuracy of the location estimates is limited by the noisy measurements. Figure 7.14 illustrates such network localization and navigation scenario.

At a given time instant, only spatial cooperation can be exploited for a static or dynamic network. According to the equivalent Fisher information matrix, the squared location error for agent k is bounded by

$$\mathbb{E}\{\|x_k - \hat{x}_k\|^2\} \geq \text{tr}\{[J_e(x)^{-1}]_{x_k}\} \tag{7.46}$$

where $[\cdot]_{x_k}$ denotes the square submatrix on the diagonal corresponding to x_k. The equivalent Fisher information matrix $J_e(x)$ consists of two parts: the localization information from anchors (i.e. block-diagonal matrices $K_k, k = 1, 2, 3$ in Figure 7.15 and that from agents' spatial cooperation (as $C_{ij}, i \neq j$ in Figure 7.15. In the absence of cooperation, $C_{ij} = 0$.

Based on the spatial cooperative localization, the technique can be extended to cooperative navigation where agents in a dynamic network

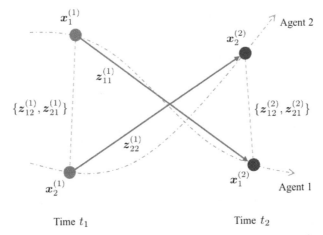

Figure 7.14   An illustration of a network with agents (blue circles) moving along the dashed trajectories. The empty ones denote those at time instant t_1 and the solid ones at time instant t_2. Intra-node measurements and inter-node measurements are denoted by green and red arrows, respectively [9].

cooperate in both the space and time domains. In each time instant, the contribution of cooperation in space is similar to spatial cooperative localization. However, another cooperation in time, exploiting intra-node measurements and mobility (dynamic) models, yields new information useful for navigation. Such information is characterized as $\mathbf{J}_e(\mathbf{x}^{1:t})$, where Figure 7.15(c) depicts the case of $t = 2$. The subsequent overall equivalent Fisher information matrix consists of two major components: cooperation in space as well as in time. The former characterizes the localization information from inter-node measurements within the entire network at each time instant, and the latter characterizes the information from intra-node measurements and mobility models at each individual agent (shown as time-domain components outside the main block-diagonal). In addition, since intra-node measurements and the mobility models for different agents are independent, the corresponding time-index matrices form a block diagonal matrix in the upper-right and lower-left quarter of the overall equivalent Fisher information matrix. These time-index components can be viewed as the temporal link that connects localization information from spatial cooperation of the previous time instant to the current one. If the temporal link is not available (i.e., time-index components are zero), the overall equivalent Fisher information matrix is block diagonal, implying that localization inference is independent from time to time. The structure of the overall equivalent Fisher information matrix

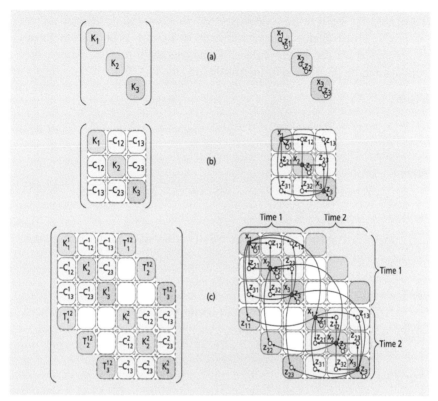

Figure 7.15 The equivalent Fisher information matrix and corresponding Bayesian networks in the 3-agent scenario: (a) non-cooperative localization (b) spatial cooperation (c) spatio-temporal cooperation [8].

for cooperative navigation allows a recursive implementation at each time instant, and realizes the information evolution of spatio-temporal cooperation in cooperative navigation.

In spatio-temporal cooperation, the agents' locational beliefs obtained through spatial cooperation are refined by information related to temporal evolution through intra-node measurements and mobility models, as depicted in Figure 7.15(c). Addition of the temporal cooperation can further increase network performance. The locational evolution is stored in the state vector $\mathbf{x}^t$, consisting of robot pose and locational derivatives at time instant t. Temporal cooperation is accomplished using mobility and intra-node measurement (likelihood) models. The former statistically describe the evolution in time of the locational states, $p(\mathbf{x}^t \mid \mathbf{x}^{t-1})$, whereas the latter statistically describe the relationship between intra-node measurements and the positional states, $p(\mathbf{z}_{self} \mid \mathbf{x}^t)$. Once again, the mechanisms to update beliefs from these models are based on the Bayes rule and marginalization. Specifically, the belief update can be performed in the following steps:

Prediction

$$p(\mathbf{x}^t \mid \mathbf{z}^{1:t-1}) = \int p(\mathbf{x}^{t-1} \mid \mathbf{z}^{1:t-1}) p(\mathbf{x}^t \mid \mathbf{x}^{t-1}) d\mathbf{x}^{t-1} \qquad (7.47)$$

Correction

$$p(\mathbf{x}^t \mid \mathbf{z}^{1:t}) = c_{norm} p(\mathbf{x}^t \mid \mathbf{z}^{1:t-1}) p(\mathbf{z}^t \mid \mathbf{x}^t) \qquad (7.48)$$

where $p(\mathbf{z}^t \mid \mathbf{x}^t) = p(\mathbf{z}^t_{self} \mid \mathbf{x}^t) p(\mathbf{z}^t_{rel} \mid \mathbf{x}^t)$.

Such cooperative mechanism can be jointly used with earlier techniques of localization and Kalman filtering.

Further Reading: The *Proceeding of the IEEE* has a special issue (vol. 106, no. 7) in 2018, which supplies useful and comprehensive information about state-of-the-art localization technology.

References

[1] D.P. Bertsekas, *Dynamic Programming*, Prentice-Hall, 1987.
[2] S. Boyd, L. Vandenberghe, *Convex Optimization*, Cambridge University Press, 2004.
[3] I. Guvenc, C.-C. Chong, "A Survey on TOA Based Wireless Localization and NLOS Mitigation Techniques", *IEEE Communications Surveys and Tutorials*, vol. 11, no. 3. pp. 107-124, 3rd Quarter, 2009.
[4] C.H. Knapp, G.C. Carter, "The Generalized Correlation Method for Estimation of Time Delay", *IEEE Tr. on Acoustics, Speech, and Signal Processing*, vol. 24, no. 4, pp. 320-327, Aug. 1976.
[5] Guoqiang Mao, Bar1s Fidan, Brian D.O. Anderson, "Wireless Sensor Network Localization Techniques", *Computer Networks*, vol. 51, no. 10, pp. 2529-2553, July 2007.
[6] H. Durant-Whyte, T. Bailey, "Simultaneous Localization and Mapping: Part I", *IEEE Robotics and Automation Magazine*, pp. 99-108, June 2006.
[7] H. Durant-Whyte, T. Bailey, "Simultaneous Localization and Mapping: Part II", *IEEE Robotics and Automation Magazine*, pp. 109-117, June 2006.

[8] M.Z. Win, A. Conti, A. Mazuelas, Y. Shen, W.M. Gifford, D. Dardari, M. Chiani, "Network Localization and Navigation via Cooperation", *IEEE Communications Magazine*, pp. 56-62, May 2011.

[9] M.Z. Win, Y. Shen, W. Dai, "A Theoretical Foundation of Network Localization and Navigation", *Proceeding of the IEEE*, vol. 106, no. 7, pp. 1136-1165, July 2018.

8

Robot Planning

In almost all application scenarios of robotics, a robot goes beyond simple response to the environment, which suggests robot planning algorithm to facilitate better actions and policies with the aid of information from sensors. This first requires the assistance of knowledge representation for further inference of collected data/information, then an appropriate planning algorithm can be developed to assist robot's intelligence of flight maneuver.

For example, in Figure 8.1, an unmanned aerial vehicle (UAV) shall fly the holes of obstacles and requires sensing the environment, SLAM, and planning algorithm, to complete the goal. The current location serves the start state of the UAV, and its goal state is successfully passing the holes of grey, orange, and finally green obstacles. The desirable planning algorithm is to find a proper flying trajectory for this UAV.

8.1 Knowledge Representation and Classic Logic

Due to actions on the physical actuators, the intelligence of a robot can not just rely on programming language and data structure, which is insufficient to derive further facts and to handle partial information. However, proper knowledge presentation and logic reasoning can serve the basis of planning for simple robot actions. Therefore, knowledge representation for robotics can be viewed as a means of representing a robot's actions and environment, as well as relating the semantics of such knowledge to its own internal components (i.e. sensors and actuators), for problem solving or mission execution through reasoning and inference [2].

Example (Blind Spot Detection): Figure 8.2 illustrates state-of-the-art blind spot detection. It is assumed that two radars are used and yellow zones indicate effective detection areas. A toy planning algorithm based on knowledge representation for an autonomous vehicle operating on the road with other

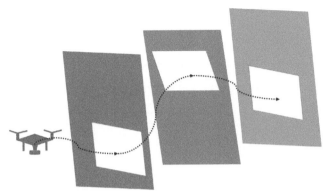

Figure 8.1   Path planning (in black dot curve) for a UAV to fly through the holes of obstacles (in grey, orange, and green).

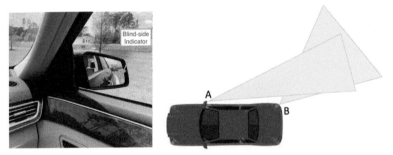

Figure 8.2   (left) State-of-the-art blind spot detection in a car (right) illustration of blind spot detection.

autonomous vehicles and human-driving vehicles can be developed using simple inference on the first-order logic, if this autonomous vehicle to turn right lane.

1. $d_A = \mathbb{I}_{detection of a metal object}$ and $d_A = \mathbb{I}_{detection of a metal object}$, where $\mathbb{I}$ is an indicator function. $D = d_A \vee d_B$, to indicate the existence of a vehicle endangering changing to right lane.
2. If $D = 1$, warning triangle on the right mirror turns orange.
3. If the driving action of this autonomous vehicle plans to change right lane, $a_{change-right} = 1$, $otherwise$ 0.
4. If $D \wedge a_{change-right} = 1$, warning triangle turns red and the action of changing right lane is prohibited. Otherwise, warning triangle remains the same.
5. Go back to step 1.

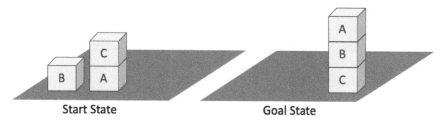

Figure 8.3 A simple robot planning case from the start state on the left to the goal state on the right.

From above example, synthesis of logic rules serves an effective way to realize the planning algorithm from the start state to goal state, on top of knowledge representation, which can be illustrated by the following simple example.

Example: Suppose there are cubic-shaped blocks on the flat ground. The blocks can be stacked and only one block can fit directly on top of another block. A robot with two arms can pick up one block and move it to another position in 3D, either on the ground or on top of another block. The robot can only hold one block at a time. As shown in Figure 8.3, the robot want to move block(s) from the start state to reach the goal state. How can we develop the robot planning based on the knowledge representation and simple (logic) operation(s)?

Solution: The actions of the robot can be represented by an operator $Move(\psi; l_i, l_j)$, which means moving block x on top of l_i to the top of l_j. In this example, ψ can be A, B, C, and l_i and l_j can be A, B, C, G (i.e. ground). The following sequence of operations can achieve the goal state from the start state:

(a) $Move(C; A, G)$
(b) $Move(B; G, C)$
(c) $Move(A; G, B)$

Of course, further detailed knowledge representation can include the concept of coordinates.

8.1.1 Bayesian Networks

To enable a robot executing human-like manipulations, ideal knowledge representation must be comprehensive to tie high-level knowledge and low-level features and/or attributes, such that the created models can be suitable for services provided by robotics. Please recall the probabilistic models suitable for dealing with uncertainty, which are useful to recognize a robot's activities and actions on its environment. *Bayesian networks* that takes advantage of probabilistic inference and graphical reasoning immediately serves this purpose.

To deal with uncertainty, *probabilistic graphical models* well serve the purpose of representation of knowledge under such uncertainty. Each vertex in graph represents a random variable and the edge between two vertices represents probabilistic dependence between corresponding random variables. Graphical models with undirected edges are generally known as *Markov random fields* or *Markov networks*. A Bayesian network corresponds to another class of graphical models, which is represented by a *directed acyclic graph* (DAG). Bayesian networks are widely applied in statistics, machine learning, and artificial intelligence. The structure of a DAG is defined as the set of vertices (or nodes) and the set of directed edges (or links).

Suppose vertices X and Y denoting two random variables. The directed edge $X \rightarrow Y$ represents the value taken random variable Y depends on the value taken by random variable X, which may suggest X influences Y. Consequently, X is referred as a *parent* of Y, and Y is referred as a *child* of X. The same concept can be generalized to define the ancestor nodes or descendant nodes. In other words, a direct arrow implies direct causal connection between two variables and thus a Bayesian network reflects a structure of relationship with uncertainty.

Prior to further advanced subjects, let us take a look at some simple Bayesian networks and their features. The probabilistic model of a Bayesian network could be written in a simple factored form, while directed edges imply direct dependence (i.e. causal relationship). The absence of an edge implies conditional independence.

Figure 8.4 illustrates two simple realizations of 3-node Bayesian networks as the foundation to study Bayesian networks. In Figure 8.4 (a), given A, B, C are conditionally independent, which gives the conditional independence as

$$p(A, B, C) = p(B \mid A)p(C \mid A)p(A) \tag{8.1}$$

An application scenario: in case A is a disease, B and C may represent conditionally independent symptoms caused by A.

Similarly for Figure 8.4 (b), A and B are marginally independent but become conditionally dependent once C is known, which gives the

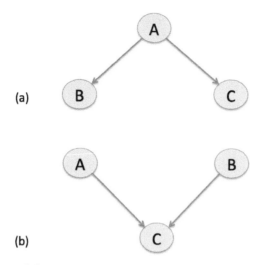

Figure 8.4 Two Realizations of 3-node Bayesian Networks.

conditional independence as

$$p(A, B, C) = p(C \mid A, B)p(A)p(B) \tag{8.2}$$

Such a scenario suggests the *explaining away* effect, that is, given C, observing A makes B less likely.

▶ **Exercise:** Ground meat purchased in the supermarket may be infected. On average, it happens once out of 600 times. A test with results positive and negative can be used. If the meat is clean, the test result will be negative in 499 out of 500 cases, and if the meat is infected, the test result will be positive in 497 out of 500 cases. Construct a Bayesian network and calculate the probability of a positive test result indeed indicating infected for meat.

The general principle of *Bayesian networks* can be understood by the following example.

Example (Security System in a Smart Home): Kawaguchi san's house in Yokohama is installed a motion sensor system to detect possible burglar entry and to automatically send short message to his mobile phone while facing burglar entry, when he is away visiting a customer in Nagoya with 3 hour commuting time. In this morning, Kawaguchi san receives the alarm message from his mobile phone. What is the probability that there is a burglar in his house? When he dials 911 to report, Kawaguchi san is told the news that there

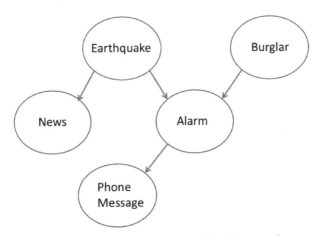

Figure 8.5   Bayesian network model of the example.

is an earth quake in Yokohama and it might trigger the alarm system. What is the probability that there is a burglar in his house?

Solution:

First of all, let us define the following variables (i.e. probabilistic events):

- B: burglar in Kawaguchi san's house
- A: alarm
- M: mobile phone alerting message
- E: earthquake in Yokohama area but not obvious in Nagoya area
- N: news regarding the earthquake

$$P(B, E, A, M, N) = P(B)P(E)P(A \mid B, E)P(M \mid A)P(N \mid E) \quad (8.3)$$

The following probabilities can be defined

- probability of burglar breaking-in: $P(B = 1) = \beta$; $P(B = 0) = 1 - \beta$; for example, $\beta = 0.001$ implies burglar rate at once in 3 years.
- probability of an earthquake happening: $P(E = 1) = \epsilon$; $P(E = 0) = 1 - \epsilon$, while E and B are independent.
- probability of alarm: The alarm system in Kawaguchi san's house has false alarm probability f to ring, say $f = 0.001$ (once in three years). Once the alarm system rings if (a) burglar entering the house with probability $\alpha_B = 0.99$ (i.e. 99% reliability to trigger the alarm if there is any burglar). (b) an earthquake to trigger the alarm with probability $\alpha_E = 0.01$.

The probabilities of A given B and E are

$$P(A = 0 \mid B = 0, E = 0) = 1 - f$$
$$P(A = 1 \mid B = 0, E = 0) = f$$
$$P(A = 0 \mid B = 1, E = 0) = (1 - f)(1 - \alpha_B)$$
$$P(A = 1 \mid B = 1, E = 0) = 1 - (1 - f)(1 - \alpha_B)$$
$$P(A = 0 \mid B = 0, E = 1) = (1 - f)(1 - \alpha_E)$$
$$P(A = 1 \mid B = 0, E = 1) = 1 - (1 - f)(1 - \alpha_E)$$
$$P(A = 0 \mid B = 1, E = 1) = (1 - f)(1 - \alpha_B)(1 - \alpha_E)$$
$$P(A = 1 \mid B = 1, E = 1) = 1 - (1 - f)(1 - \alpha_B)(1 - \alpha_E)$$

It is also reasonable to assume

- $P(M = 1 \mid A = 1) = 0$, that is, there will be no mobile phone alerting message if no alarm.
- $P(N = 1 \mid E = 0) = 0$, the news regarding the earthquake is reliable.
- The event of burglar and the event of earthquake are independent.

When $M = 1$, we know that there exists an alarm, $A = 1$. According to the Bayes theorem, the posteriori probability of B and E is

$$P(B, E \mid A = 1) = \frac{P(A = 1 \mid B, E)P(B, E)}{P(A = 1)} \tag{8.4}$$

The probability that a burglar was in Kawaguchi san's house can be obtained by marginalizing over E, to give numerical results as

$$P(B = 1 \mid A = 1) = 0.505$$
$$P(B = 0 \mid A = 0) = 0.495$$

Once getting mobile alerting message, it is really hard to judge whether there is a burglar or not. However, when $E = 1$ (i.e. learning from the news), the posteriori probability of B is

$$P(B = 1 \mid E = 1, A = 1) = 0.92$$
$$P(B = 0 \mid E = 1, A = 0) = 0.08$$

It is a much smaller chance that Kawaguchi san should go home for burglar breaking in.

Bayesian network model is actually a *graphical model* combining graph theory and probability theory to suggest a general framework to represent models of interactive variables, and are widely adopted into problems in AI and robotics. Each node in a graph represents a random variable (or a set of random variables). The pattern of edges in the graph represents the qualitative dependencies between/among the variables, while such quantitative dependencies between variables connected by edges are specified via (non-negative) *potential functions*. Two common forms of graphical models are *directed graphical model* and *undirected graphical model*. For directed graphs, the potential function turns out to be the conditional probability of the node given its parents.

Let $\mathcal{G}(\mathcal{V}, \mathcal{E})$ be a directed acyclic graph, where $\mathcal{V}$ denotes the nodes and $\mathcal{E}$ denotes the edges of the graph. $\{X_v : v \in \mathcal{V}\}$ denotes a collection of random variables indexed by the nodes of the graph. For each node $v \in \mathcal{V}$, π_v denotes the subset of indices of its parents. X_{π_v} denotes the random vector indexed by the parent of v. Given a collection of kernels, $\{k(x_v \mid x_{\pi_v}) : v \in \mathcal{V}\}$, with normalization to one, we can define a joint probability distribution

$$p(x_v) = \prod_{v \in \mathcal{V}} k(x_v \mid x_{\pi_v}) \tag{8.5}$$

Since this joint probability distribution has $\{k(x_v \mid x_{\pi_v}) : v \in \mathcal{V}\}$ as its conditionals, we can write $k(x_v \mid x_{\pi_v}) = p(x_v \mid x_{\pi_v})$, while $p(x_v \mid x_{\pi_v})$ is the local conditional probability associated with node v.

For undirected graphs, the basic subsets are called *cliques* of the graph, that is, subset of nodes that are completely connected. For a given clique C, $\psi_C(x_C)$ denotes a general potential function assigning a positive real number to each configuration x_C. We have

$$p(x) = \frac{1}{Z} \prod_{C \in \mathcal{C}} \psi_C(x_C) \tag{8.6}$$

where $\mathcal{C}$ is the set of cliques associated with the graph and Z is the normalization factor to ensure $\sum_x p(x) = 1$. Please note that $p(x_v \mid x_{\pi_v})$ is a perfect example as a potential function.

In the following, the probability inference will be briefly introduced over the graphical models. Let (E, F) be a partitioning of the node indices of a graphical model into disjoint subsets, such that (X_E, X_F) indicates a partitioning of the random variables associated with the graph. For two common inference problems of interest:

- marginal probabilities:

$$p(x_E) = \sum_{x_F} p(x_E, x_F) \tag{8.7}$$

- *maximum a posteriori* (MAP) probabilities:

$$p^*(x_E) = \max_{x_F} p(x_E, x_F) \tag{8.8}$$

Based on these computations, further results of interest can be developed. For example, the conditional probability

$$p(x_F \mid x_E) = \frac{p(x_E, x_F)}{\sum_{x_F} p(x_E, x_F)} \tag{8.9}$$

Similarly, supposing (E, F, H) to be a partition of the node indices, the conditional probability can be computed as

$$p(x_F \mid x_E) = \frac{p(x_E, x_F)}{\sum_{x_F} p(x_E, x_F)} = \frac{\sum_{x_H} p(x_E, x_F, x_H)}{\sum_{x_F, x_H} p(x_E, x_F, x_H)} \tag{8.10}$$

With probabilistic graphical models, edges would describe the likelihoods of certain variables as nodes causing others to occur. Using above principles of inference, many algorithms can be therefore developed for the cases of interest.

▶ **Exercise:** Dr. Minkowski is 92 years old and lives alone. A nursing robot is taking care of him. To ensure balance food and nutrition, the nursing robot must make sure two or three kinds of fruits of enough amount available at home. Too many kinds of fruits may have a challenge to ensure fruits fresh. The smart refrigerator notifies the nursing robot that only strawberry inside the refrigerator to order new fruit delivery, with probability of miss detection to be 3%. The nursing robot scans the kitchen as Figure 8.6. In addition to strawberry inside refrigerator, sufficient number of oranges are identified in the red marked window, with probability of incorrect identification (i.e. not enough fruit other than strawberry) to be 5%. There are fruits outside the kitchen with only 10% chances. If the probability of one kind of fruit or no fruit at home is higher than 95%, the nursing robot should order new fruit delivery. Please derive the decision graph for this nursing robot to order a new fruit delivery or not, and the associated decision mechanism of ordering food delivery.

Figure 8.6   Scanned Image and Object Identification of Kitchen; the bottle of detergent that is not eatable is marked as blue; potential but unlikely containers (oven, microwave oven, rice cooker, coffee maker) are marked by green; identified fruits are marked in red while the refrigerator known to have strawberry inside.

▶ **Exercise:** It is always challenging to design an autonomous driving vehicle (AV), particularly with human driving around. Suppose an AV is driving behind a huge truck on the right lane of a two-lane street. This huge truck blocks all visionary devices (LIDAR, mmWave RADAR, camera) on the AV. If there is a slow car in front of the truck, the driver may brake or change lane. If there is a car braking in front of the truck, the truck has to brake. This AV knows no vehicle closely behind. Please develop the Bayesian network for this AV's action to brake or change lane, while observing the braking light of the truck in front.

8.1.2 Semantic Representation

Graphs are useful in knowledge representation. Instead of probabilistic graphs, *semantic graphs* whose nodes and edges describe semantic concepts and details between entities as observed can be developed. For example, spatial concept can be described by semantic graphs, where nodes can describe objects within a scene, and edges describe commonality or contextual

Figure 8.7 Red rectangles indicate the objects related to robot's movement; yellow rectangles indicate the doors that might not be useful for robot's movement; blue rectangles indicate objects non-related to plants; green rectangles indicate plants.

relationships between objects in terms of position, such as one object may be on top of another object in Figure 8.3. Temporal relationship (i.e. two or more events related by time) can be also embodied by semantic graphs.

Exercise: As Figure 8.7, a robot is trying to take proper movements to water the plant in a living room. Please develop the semantic graph to properly represent the knowledge toward this goal.

8.2 Discrete Planning

Planning and problem solving are highly similar subjects in artificial intelligence. When A^* algorithm was introduced in Chapter 2 Search Algorithms, it is known useful to planning-type problem solving. In this section, general discrete planning will be oriented. Again, the concept of state space will be used to define the problem. Each distinct situation for the world is called a *state*, denoted by x, and the set of all possible states is called a *state space*, $\mathcal{X}$. The mathematical meaning of discrete implies countable, and usually finite in planning. The world can be transformed by *actions* selected by the planner.

Each selected action a, when is applied to the current state x, produces a new state x', which is specified by a *state transition function*, Φ.

$$x' = \Phi(x, a) \tag{8.11}$$

Remark: When we define the state, all relevant information should be included, but any irrelevant information should not be included to avoid unnecessary complexity.

Let $\mathcal{A}_x$ denote the action space for state x, as the set of all possible actions that can be applied to state x. For distinct $x, x' \in \mathcal{X}$, $\mathcal{A}_x, \mathcal{A}_{x'}$ are not necessarily disjoint, since the same action may be applicable to different or even all states. Consequently, the set $\mathcal{A}$ is defined to represent all possible actions over all states.

$$\mathcal{A} = \bigcup_{x \in \mathcal{X}} \mathcal{A}_x \tag{8.12}$$

By defining a set of goal states, X_G, a planning algorithm is to identify a sequence of actions (i.e. a policy) to transform from the initial state, x_I, to some state in X_G. After definitions of the planning problem, it is common to construct a directed state transition graph. The set of vertices means the state space. A directed edge from x to x' exists if and only if there is an action $a \in \mathcal{A}_x$ such that $x' = \Phi(x, a)$.

Example: Suppose a mobile robot can move up, down, left, and right in Figure 8.8, but not into the obstacle (i.e. black tiles). The states represent the positions. Actions include four possible movement but less in some states due to obstacles. Suppose the green tile is the starting position of movement as the initial state, and the red tile is the goal state to terminate.

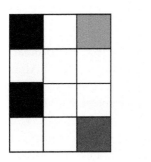 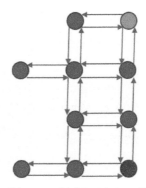

Figure 8.8   (left) Tiles that a mobile robot can possibly move (right) state transition graph.

It is straightforward to plan the movement of robot, as a typical graphical algorithm, and graphical searching algorithms generally work for planning. Any planning algorithm follows the following principle to construct the search graph $\mathcal{G}(V, E)$:

1. Initialize: Initiate $\mathcal{G}(V, E)$ with empty E and some starting states for V. For forward search, $V = \{x_I\}$. For backward search, $V = \{x_G\}$. The search graph grows step-by-step to reveal more state transition graph.
2. Select vertex: Choose a vertex $n_{exp} \in V$ for expansion, with x_{exp} as its state.
3. Apply an action: A new state, x_{new} is obtained by $x_{new} = \Phi(x, a)$ for $a \in \mathcal{A}_x$ in forward search, or $x = \Phi(x_{new}, a)$ for $a \in \mathcal{A}_{x_{new}}$ in backward search.
4. Insert a directed edge to the search graph: Once certain algorithm-specific conditions are satisfied, generate a directed edge from x to x_{new} for forward search, or from x_{new} to x for backward search.
5. Check goal: Determine whether $\mathcal{G}$ creates a path from x_I to x_G.
6. Iterate: Repeat (go back step 2) until either a solution is found or a termination condition is met.

▶ **Exercise:** For Figure 8.8, please find a plan for the mobile robot moving from the green tile to the red tile.

Of course, the purpose of planning algorithms is not just to find a plan, and a good or optimal plan is desired, which leads to the *optimal planning*. The optimality is subject to certain criterion to measure, with typical cases such as time, distance, energy consumption, etc., which lead to the introduction of *cost* functions. Consequently, the optimal planning minimizes the cost or maximize the *reward*. Suppose such cost functions are known and static with time for the time being. Similar to the concept of finite horizon, denote δ_K as the K-step plan consisting of a sequence of actions, $a_1, a_2, \cdots, a_K$ with index indicating the step. When δ_K and x_I are given, a sequence of states can be derived according to the state transition function Φ. Initially for $x_1 = x_I$, then $x_{k+1} = \Phi(x_k, a_k)$.

Given the additive cost functional (analogous to the concept of length), L, the formulation of *discrete fixed-length optimal planning* involves

- $\mathcal{X}, \mathcal{A}_x, \Phi, x_I, X_G$ as earlier definitions, with current interest of finite $\mathcal{X}$.
- K stages in a plan corresponding to actions $a_1, \cdots, a_K$. States are also labeled by the stages, say x_{k+1} after action $a_k, k = 1, \cdots, K$. x_{K+1} therefore represents the final state.

- The cost functional is

$$L(\delta_K) = \sum_{k=1}^{K} l(x_k, a_k) + l_{K+1}(x_{K+1}) \tag{8.13}$$

where $l(x_k, a_k) \in \mathbb{R}^+, \forall x_k \in \mathcal{X}, a_k \in \mathcal{A}_{x_k}; l_{K+1}(x_{K+1} = 0$ if $x_{K+1} \in X_G$ and $l_{K+1}(x_{K+1} = \infty$ otherwise.

The optimal planning can be obtained by

$$\min_{a_1, \cdots, a_K} \left\{ \sum_{k=1}^{K} l(x_k, a_k) + l_{K+1}(x_{K+1}) \right\} \tag{8.14}$$

which is a typical dynamic programming or backward search problem by value iterations (say, Dijktra algorithm) in Chapter 2.

8.3 Planning and Navigation of An Autonomous Mobile Robot

In this section, following the localization in previous chapter, the focus is to execute the planning and navigation of an autonomous mobile robot (AMR), which can be an autonomous guided vehicle in a smart factory or a robot to automatically mow the lawn by execute automated movement to complete its goal.

Given a reference coordinate system (i.e. a map) and a destination, *path planning* identifies a trajectory for an AMR to execute automate movements reaching the destination and avoiding the obstacles, which can be viewed as the strategic problem-solving. Let n be the time-index, and the AMR have the map $\mathcal{M}_n$ and belief $\mathbb{B}_n$ at time n. The first step of path planning is to determine the *configuration space* for robot's location and/or pose. Suppose a robot has κ degrees of freedom (DOF), which means that every state or configuration of this robot requires κ parameters to quantitatively describe. Configuration space may also contain information regarding a grid available to move or having an obstacle in certain position associated with a map. Path planning intends to connect the starting point and the destination for the robot's movement over a road network, which can typically be represented by a graph. In many cases, if a robot is guided by visionary devices, visibility graph for a polygonal configuration space consists of edges jointing all pairs of vertices that can be observed while removing the edges toward obstacles.

The subsequent task of path planning is therefore to find the shortest path in the first two chapters of this book.

Above configuration space appears well defined, but the real application scenarios may not. Consequently, we have to

- decompose the operating environment into connected regions called *cells*.
- determine the available/open and adjacent cells to construct a *connectivity graph*.
- determine the starting point and destination and search for a path in the connectivity graph.
- compute a sequence of movements based on the κ parameters of a dynamic systems (i.e. the robot).

Above procedure is called *cell decomposition path planning*, while almost all examples and exercises in this book consider this way and ideally adopt square grids as cells. Such a method is called *exact cell decomposition*. If the decomposition results in an approximation of the actual map, it is known as *approximate cell decomposition* and serves a popular method for mobile robot planning. For example, the *grassfire algorithm* that will be used later in this section is an efficient and simple-to-implement technique for finding an appropriate route in such approximate fixed-size cell arrays of the environment. The algorithm simply employs wavefront expansion from the goal position outward, marking for each cell its distance to the destination cell. This process continues until the cell corresponding to the initial robot position is reached. At this point, the path planner can estimate the robot?s distance to the destination (or goal state) as well as discovering a specific solution trajectory by simply linking together cells that are adjacent and always closer to the goal.

8.3.1 Illustrative Example for Planning and Navigation

In the following, we consider an illustration for an AMR using RL and planning algorithm to complete its mission (or reach its goal). Let a cleaning robot to work on a large floor consisting of a big number of square tiles as shown in Figure 8.9 corresponding to a grid-based map. Each robot can move up, down, left, and right, in one unit of time. The cleaning task of one tile/grid can be done within the same time unit. These automated cleaning robots share the same mission (i.e. to clean the entire floor) but each of them executes on own intelligence without any centralized controller to manage their actions,

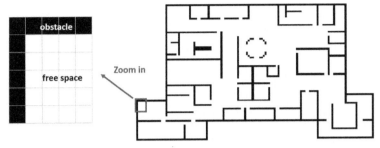

Figure 8.9   Floor plan of the cleaning area, where the area consists of 6760 free space grids and 1227 obstacle grids.

as a collaborative MAS. To make this example more meaningful in diverse application scenarios, we assume

- The size and shape of the target area is time-invariant but unknown to robots (i.e. agents). In other words, the map of the target area is not available to agents.
- Each agent does not know its location at the beginning and must explore to establish its private reference (i.e. own but incomplete map of target area).
- Each agent equips appropriate sensors and localization algorithm to tell each tile is to-be-cleaned, being cleaned before, and a block. Each agent executes its own learning and decision, which will be modeled as reinforcement learning.
- The time to complete the mission of cleaning the entire (or certain percentage of) floor is used as the performance index of such AI (single-agent or multi-agent) system.

Each agent equips sensors to precisely observe neighboring 4 grids (upper, lower, left, and right), to precisely move and clean in one time unit, but without knowing the floor map. The agent must represent the environment by occupancy grid map, which means the robot must generate own private reference as a solution of localization.

8.3.2 Reinforcement Learning Formulation

Based on the system model with uncertainty to each agent (i.e. robot), it is appropriate to employ reinforcement learning to represent each agent's behavior. The *target area* is represented in grid-based map. Each grid represents a square tile of unit length at each side, uniquely labelled by $g_{p,q}, p, q \in$

$\mathbb{N}$, which directly indicates its geometric position (p, q) on the grid map. Every grid can belong to one of these types:

- Obstacles: A fully occupied grid that is not able to let the robot traverse. Obstacle grids can be represented by $\mathcal{M}_{obs}$.
- Unvisited (uncleaned): A grid that is covered with dirt but have not yet been cleaned. The set of unvisited grids is $\mathcal{M}_X$.
- Visited (cleaned): A grid that has been cleaned and doesn't need to be visited again. The set of visited grids is $\mathcal{M}_O$. (Note that unvisited grids and visited grids are both *free space*, which is denoted by $\mathcal{M}_{free}$.)

The grid-based map of a target area is denoted by $\mathcal{M}$, which contains information about the size and shape of the target area, the label of grids, and the type of grids. Agent is assumed to follow a discrete time schedule $t \in \{0, 1, 2, \dots\}$. It is able to move from the current grid to the center of one of the four adjacent grids at each time instance t. During time slot $(t, t+1)$, it cleans up the current grid it is occupying. The agent based on its observations (a grid is obstacle or not) and experience (whether it has visited the grid) to establish and update its own map, relative to the agent's initial location without knowing the true map. Such self-constructed map actually forms the *private reference*. Therefore, agent u_i's private reference at time t is denoted by $\mathcal{M}_t^i$.

Let the state of the agent u_i be defined by its current location $g_{p,q}^i$. For example, given that at time t the agent u_i is on grid $g_{p,q}^i$, we use $y_t^i = g_{p,q}^i$ to represent its state. Note that the superscript in $g_{p,q}^i$ indicates t using its private reference. For any state y_t^i, the possible action can be determined based on earlier assumptions. In other words, it knows which neighbor grids are traversable. With this assumption and letting $\mathcal{N}(y_t^i)$ be the collection of four neighbor grids, we thus can define the action set for state y_t^i to be

$$A(y_t^i) = \{y | y \in \mathcal{N}(y_t^i) \cap y \in \mathcal{M}_{free}\}. \tag{8.15}$$

Since the agent is assumed to be able to accurately control its motion, the state transition probability $p(y_{t+1}^i | y_t^i, A_t)$ is known. Suppose the previous state is $y_t^i = g_{j,k}^i$, the new state is certain to be

$$y_{t+1}^i = \begin{cases} g_{j,k}^i & \text{if } A_t = stay \\ g_{j,k+1}^i & \text{if } A_t = forward \\ g_{j,k-1}^i & \text{if } A_t = back \\ g_{j-1,k}^i & \text{if } A_t = left \\ g_{j+1,k}^i & \text{if } A_t = right. \end{cases} \tag{8.16}$$

Every time the agent takes an action, besides transitioning into next state, it also receives a real-valued reward R_{t+1}, while the design of reward function is a critical part in reinforcement learning. Given the agent's state $y^i_{t+1} = g^i_{p,q}$, its actual location can be obtained by a shift operator T_i. For instance, if u_i's initial location with respect to realistic map is $g_{j,k}$, the operator T_i yields $T_i(g^i_{p,q}) = g_{p+j,q+k}$, while the agent regards its initial location as the origin point of $\mathcal{M}^i$. Hence, for $T_i(y^1_{t+1}) = g_{a,b}$, in order to encourage the robot explore the map, we set the reward structure to be

$$R_{t+1} = \begin{cases} R^+ & \text{if } g_{a,b} \text{ has not been cleaned} \\ R^- & \text{otherwise.} \end{cases} \tag{8.17}$$

R^+ can be expressed as $R_{good} - E_1$. R_{good} is a positive value that encourages agent to gradually complete the cleaning task. E_1 represents the cost of moving from one grid to another. When a grid has been cleaned, $R^- = -E_1$ as punishment.

It is ready to select baseline reinforcement learning for an agent. Due to noisy sensing, the value on reward map $\widetilde{R}(g)$ suffers error with probability p_e.

$$P\{\widetilde{R}(g) = R^-|g \text{ dirty}\} = P\{\widetilde{R}(g) = R^+|g \text{ cleaned}\} = p_e \tag{8.18}$$

$$P\{\widetilde{R}(g) = R^-|g \text{ cleaned}\} = P\{\widetilde{R}(g) = R^+|g \text{ dirty}\} = 1 - p_e \tag{8.19}$$

Since error happens independently at each time when agent looks up the reward map, error occurs independently in $R(g), \forall g$ at any time t. As the true state is hidden, the agent has to estimate the underlying state, the estimated state is referred to as *belief states*. We use b^i_t to represent it. The agent applies Q-learning to complete floor cleaning task subject to hidden information, with ϵ-greedy for exploration due to unknown floor map, while the final algorithm is summarized as follows.exhaustive

1) Agent is randomly deployed at a free space grid which is defined as coordinate $(0,0)$ in its private reference $\mathcal{M}^i_t, t = 0$.
2) Perceives 4 surrounding grids. For all a in action set $\mathcal{A}(b^i_t)$, initialize $Q(b^i_t, a) = Q_0$ if $Q(b^i_t, a)$ has not been defined. Q_0 is just an initial value that could be set at any value.
3) Calculate action-value function using $\widetilde{R}_{t+1}$ in reward map. The tilde over $\widetilde{Q}(b, a)$ indicates that it is not the real action-value but the estimated one. $\forall a \in \mathcal{A}(b^i_t)$,

$$\widetilde{Q}(b^i_t, a) \leftarrow Q(b^i_t, a) + \alpha[\widetilde{R}_{t+1} + \arg\max_{a'} Q(b^i_{t+1}, a') - Q(b^i_t, a)] \tag{8.20}$$

where $\widetilde{R}_{t+1}$ follows Equation (8.17) but subject to error as mentioned in Equation (8.18); y_{t+1}^i follows Equation (8.16).

1. Let the optimal action be $a^* = \arg\max_a \widetilde{Q}(b_t^i, a)$. Choose action A_t following ϵ-greedy policy, that is

$$A_t = \begin{cases} a^* & \text{with probability } 1 - \epsilon \\ a \neq a^* & \text{with probability } \frac{\epsilon}{|\mathcal{A}(b_t^i)|-1}. \end{cases} \qquad (8.21)$$

2. Operate action A_t, transits to state b_{t+1}^i, and receive reward R_{t+1}. Update the action-value function

$$Q(b_t^i, a) \leftarrow Q(b_t^i, a) + \alpha[R_{t+1} + \arg\max_{a'} Q(b_{t+1}^i, a') - Q(b_t^i, a)] \quad (8.22)$$

3. $t \leftarrow t + 1$. If all rewards on reward map is R^-, meaning that all grids have been cleaned, terminate. Otherwise, go back to step (ii).

After experimenting a few basic types of reinforcement learning for each agent's actions using sensor data, Q-learning to estimate the action-value functions turns out to be effective and we will use as baseline learning by incorporating further techniques, while temporal difference learning in n steps (TD-n) suffers from no global map information.

Purely relying on reinforcement learning is very ineffective to complete the task for a large area without floor map (i.e. public reference among collaborative agents). To tackle this dilemma, two paradigms for planning on private reward map will be introduced for the purpose of localization and planning. The planning algorithms are imposed on the agent's behavior policy. Hence the agent adopts the original behavior policy (e.g. ϵ-greedy) and a planning policy interchangeably. The first planning algorithm is called *fixed depth planning*, in which agent retrieves limited part of the information in its reward map at every decision epoch. The second paradigm imposes switching conditions for agent to start exhaustive planning on reward map, and thus named as *conditional exhaustive planning*.

8.3.3 Fixed Length Planning

Fixed depth planning (FDP) follows a straightforward way to exploit reward map. It checks for any existence of positive rewards in the reward map before using ϵ-greedy policy. That is, in spite of deciding action according to the action value, the agent first examines all grids within Manhattan distance d

and goes toward the nearest grid that has positive reward. Manhattan distance between agent's current location $g_{a,b}^i$ and any grid $g_{p,q}^i$ is defined as

$$md(g_{a,b}^i, g_{p,q}^i) = |a - p| + |b - q| \qquad (8.23)$$

Consequently, fixed length planning consolidated with Q-learning can be summarized by augmenting an extra step 2a) between 2) and 3) in Section 8.3.2:

2a) Let the set of grids that are reachable from b_t^i in m steps be denoted by $\mathcal{N}^m(b_t^i)$. Choose A_t such that the proceeding believe state b' is getting the agent closer to a grid of positive reward (limited to Manhattan distance d). Tie breaks evenly.
If within distance d no positive reward exists, proceed to step 3). Otherwise go to step 5).

8.3.4 Conditional Exhaustive Planning

In the scenario of floor cleaning tasks, as well as other tasks that cope with unknown environment, trade off between exploration (in the environment) and exploitation (accomplish the agent's major task) solidly exists. Fixed depth planning is a typical case of constantly trying to exploit, trading off task completion time over computation cost. When the robot explores, it has a more thorough understanding of the environment. More importantly, exploration helps to establish a private reference and private reward map which become valuable resource that can be provided for others and for its future decision. But a fully exploration behavior is very likely to waste additional unnecessary energy and time. It is therefore desirable to identify a reasonable condition to determine when the planning should start. A straightforward path planning algorithm called *grassfire algorithm*, that exhaustively searches for positive rewards on reward map, is introduced to complete the concept of *conditional exhaustive planning* (CEP).

Conditions for Adopting Planning: Since exploration and planning are both likely to benefit the agent's task and thus hard to seek a balance, we consider the undesirable situations that one of them, either exploration or planning has worse performance. For instance, the robot just finishes cleaning up the entire room or to the end of a hall way. If it keeps strolling around that same area, the exploration progress is certainly paused. More intuitively, the agent is prohibited to proceed exploration for a short period of time after it just cleaned up a *block*.

A block can be regarded as a confined area. We let the size of a block to be N_B. In our way of map representation, i.e. grid map consists of square of unit length, an area of size N_B can be regarded as N_B consecutive girds. So, whenever the robot moves inside the block more than N_B steps, it will definitely visit some of the grids at least twice.

Planning Algorithm (Grassfire Algorithm): In the planning, the agent's objective is to clean the unexplored area according to the best of its current knowledge. There are several way to achieve this, either to plan a path that could cover all uncleaned grid, or search its private reference and randomly select one grid as the next destination. However, a trade off between exploration and exploitation (i.e. planning to clean the entire area in private reference) is critical. To avoid repeating the same mistakes, such as greedy exploitation in local planning, we decide to let the agent plan moderately. Hence, the agent is supposed to make a simple plan just for escaping the block. The most efficient way to escape the block is to search for and go to the nearest uncleaned grid. Grassfire algorithm, a breadth-search first method on a grid-based graph, can effectively search for the goal on a graph and is able to construct a path from starting point to the goal. Therefore, the Grassfire algorithm can serve the planning algorithm that helps the robot navigate to the nearest uncleaned grid. Since the Grassfire algorithm does not stop searching until finding an unvisited grid, it is a kind of exhaustive search.

Grassfire algorithm can be implemented as a grid-based breadth-first search (BFS) algorithm. BFS starts from a point and expands to neighbor grids that have not been searched yet. Searching procedure ends either when the goal is reached or when expansion is not allowed, i.e. all grids are examined. A somehow contrasting method is depth-first search (DFS). BFS and DFS differ majorly in how they decide the priorities of node when expanding searching. DFS chooses a node to expand at one time and stops until the end of a branch. If the goal is not found, go back to the lowest layer that has unexpanded node and start searching another branch. Instead of searching until the end of a branch, BFS expands to search every neighbor node that is directly connected with the currently expanding node. All explored neighbor nodes will be queued as future candidate expanding nodes and will be removed from the queue once all its neighbor is explored. The candidate expanding nodes are queueing in order and will serve in a first-in-first-out manner. Therefore, the shortest distance between any node and the original node can be computed. BFS applying on a grid-based graph is also known as grassfire algorithm.

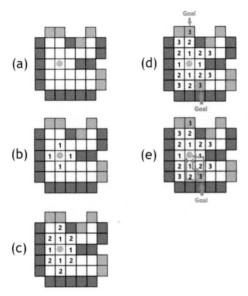

Figure 8.10   Grassfire algorithm: breadth-first search on grid-based map, where the yellow circle represents the current position from which we try to find a nearest goal, i.e. green grids. Please refer to Section 8.3.4.

The computation complexity will be $\mathcal{O}(|V|)$ where $|V|$ is the number of nodes (i.e. grids) in the graph. Figure 8.10 shows the process.

1. Set the starting position.
2. Start from current location to expand. Explore all neighbors and check whether there is any green grid.
3. Since no goal is found, let the previous explored grids become the expanded ones. An explored grid will not be searched again.
4. Once goals are found, although here we show two goals, the expansion process in fact undergoes in order. If it expands from the grid labelled with 2 at the top, it will find the upper goal first. On the contrary, if it expands from one of those two grids labelled with 2 on the lower right corner, it finds the goal locating at the lower half.
5. Suppose we find the lower right hand side goal. To construct a path from origin to destination, we first trace a path backwards and reverse it. That is, construct a path from the goal - simply move towards the neighbor with the smallest distance value with breaking ties arbitrarily until arriving the origin. Then reverse the path. Figure 8.10 shows three shortest paths in orange line.

Learning N_B**:** There is no rule to define the size of a block N_B, and the proper size of the block could be relative to the pattern in the environment. Therefore, we employ a reinforcement learning scheme again for the agent to dynamically select the block size. In order to evaluate how good a certain N_B value is, first we set up a range for N_B, that is $\{N_B | N_B \in \mathbb{N}, A \le N_B \le B\}$. All value is initialize as zero. $V(N_B) = 0, \forall N_B$ In the beginning, robot randomly selects a value to define the block size. Once it changes from exploration mode to planning mode, it records all rewards and cost it collects along the path from s_P to s_G. But because s_G is the closest unvisited grid, the robot will not receive any positive reward for sure. The total reward and cost is

$$Cost(s_P, s_G) = md(s_P, s_G) \times E_1. \tag{8.24}$$

Starting from the goal s_G, robot also records the rewards and cost it receives in the next N_B steps. These returns collected in the time period since arriving s_G to N_B steps later can be written as

$$G_{t(s_G):t(s_G)+N_B} = \sum_{i=t(s_G)+1}^{t(s_G)+N_B} R_i \tag{8.25}$$

where $t(s_G)$ represents the smallest time index when reaching s_G count from when planning started.

Having (8.24) and (8.25), the robot can online update $V(N_B)$ by

$$V(N_B) \leftarrow V(N_B) - \alpha[G_{t(s_G):t(s_G)+N_B} - Cost(s_P, s_G) - V(N_B)]. \tag{8.26}$$

Every time any value function of N_B is modified, robot chooses N_B with the highest value function since that N_B is possible to bring more profits based on its experience. Figure 8.11 indicates remarkable efficiency of the Q-learning incorporating with planning and localization, even very simple or straightforward planning algorithm, which suggests useful rule of thumb to design AI mechanism for a robot. Of course, such algorithms will be used as baseline to consider wireless communications in a multi-robot system in Chapter 10.

■ **Exercise:** A robot is going to clean an office with layout as Figure 8.12. However, this robot has no prior knowledge of office layout. The robot equips sensors to perfectly sense the status of 4 immediate neighboring grids (left, right, up, down) and move accordingly. The black grids are prohibited to enter (i.e. wall). The robot has memory to store the grids being visited and being

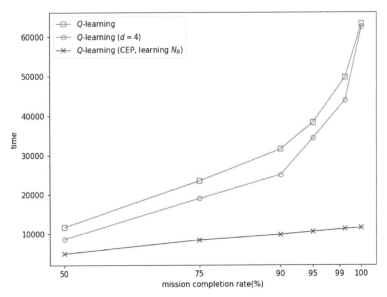

Figure 8.11 Private reference planning with learning block size, where Parameters are fixed at $\epsilon = 0.1, \alpha = 0.1, \gamma = 0.9, R^+ = 1, R^- = -0.5, Q_0 = 1$ [5].

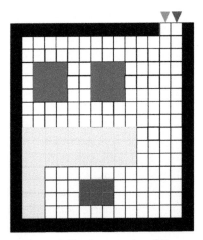

Figure 8.12 Layout of an Office.

sensed, with relative position to the entry point to form its own reference (i.e. map). Suppose all colored grids are not accessible. Please develop the RL algorithm and simple planning algorithm based on the memory to visit all white grids and leave from the entrance (indicated by green and red

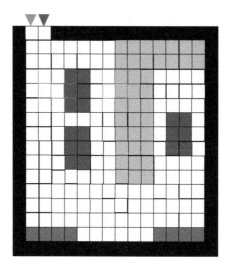

Figure 8.13 Layout of another Office.

triangles) of the office. How many steps does the robot take based on your programming?

■ **Exercise:** Please use the same algorithm and codes of previous exercise to repeat as Figure 8.13. Is there any difference? If so, how to improve for universal layout?

Further Reading: The basic AI knowledge of robot planning can be found in [1]. More in-depth treatments of Bayesian network and decision graphs can be found in [3]. The basic planning for an agent or robot can be found in [1, 5]. However, more in-depth knowledge of robot planning and its mechanics can be found in [4].

References

[1] Stuart Russell, Peter Norvig, *Artificial Intelligence: A Modern Approach*, 3rd edition, Prentice-Hall, 2010.
[2] D. Paulius, Y. Sun, "A Survey of Knowledge Representation in Service Robots", *Robotics and Autonomous Systems*, vol. 118, pp. 13-30, 2019.
[3] F.V. Jensen, T.D. Nielsen, *Bayesian Networks and Decision Graphs*, 2nd edition, Springer, 2007.

[4] K.M. Lynch, F.C. Park, *Modern Robotics: Mechanics, Planning, and Control*, Cambridge University Press, 2017.

[5] K.-C. Chen, H.-M. Hung, "Wireless Robotic Communication for Collaborative Multi-Agent Systems", *IEEE International Conference on Communications*, 2019.

9

Multi-Modal Data Fusion

A state-of-the-art robot equips multiple kinds of sensors to understand the environment for actions of better quality and reliability. The consequent emerging technology to fuse multiple (sensor) information, *multi-modal data fusion*, is required to facilitate machine intelligence. Figure 9.1 depicts the sensing-decision mechanism in a typical autonomous mobile robot. There are multiple kinds of sensors to sense the environment, together with information from cloud/edge management, multi-modal data fusion is executed to enable robot's decisions and actions.

9.1 Computer Vision

Vision might be the most important sensing capability for human beings, with majority of brain functioning for vision. Computer vision is a technology allowing a computer or a computing-based agent to obtain high-level understanding from (digital or digitized) images or videos. As a matter of fact, a video consists of multiple image frames and their deviations in a second, where human can have "continuous" vision on the video. Though an image or a video is 2-D in principle, 3-D understanding is possible. Computer vision or later machine vision is composed of two components:

- A sensing device captures as many details from the target environment to form an image as possible. Human eyes capture light coming through the iris and project it to the retina, where specialized cells transmit information to the brain through neurons. A camera that is most common in computer vision and machine vision captures images in a similar way by forming pixels, then transmit the information to the computer or vision processing unit. State-of-the-art cameras are better than humans as they can see infrared, see farther away or with much more precision.

241

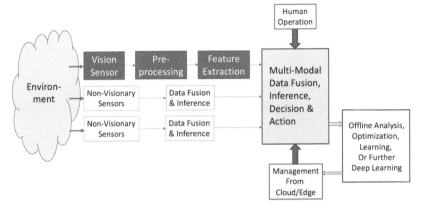

Figure 9.1   The Mechanism of Multi-Modal Sensing for A Robot's Decisions and Actions

Figure 9.2   Six photos for "I am not a robot test".

- An interpreting device processes or computes the information and extract high-level understanding from the image. The human brain works perfectly in multiple steps and in different regions of the brain. From this end, computer vision still lags behind human vision perception, but can deals with large number of images via deep learning for certain applications.

Example (I am not a robot test): In Figure 9.2, there are six photos (A-F) and please select the photos of residential house(es) with garage door in front. It is rather easy for a human to select but pretty challenging for a computer or software robot to determine in a short time.

9.1.1 Basics of Computer Vision

The origin of computer vision can be traced back to a summer project at MIT in 1966, but now involves multi-disciplinary knowledge ranged from computer science, engineering, mathematics, to psychology. Computer vision is hard because it is not just to develop pixels and 3D modeling, but also interpret the meaning. Therefore, computer (or animal) vision relies on the sensing devices and the interpreting devices. David H. Hubel and Torsten N. Wiesel, the laureates of 1981 Nobel Physiology and Medicine Prize, pioneer animal (and human) vision and brain functions.

Three primary colors, red, green, blue, are usually used to represent the color in computer vision. White Balance is a process to adjust the image data received by sensors to properly render neutral colors (white, black, and gray levels), which is performed automatically in state-of-the-art digital cameras with proper filtering. A pixel is the basic element of an image, typically in squares. A color image pixel in the RGB model has quantized intensity values from 0 to 255 in each of red, green, and blue channels. A 3-D tensor can represent a color images (x and width, y and length, red-value, green-value, blue-value).

Example: Suppose there is an image, which consists of $N_x N_y$ pixels that has width Δ_x and length Δ_y, as shown in Figure 9.3. We may use the tensor $(x, y, q_{red}, q_{green}, q_{blue})$ to represent this pixel in color. After this basic step, further comprehension is possible in computer vision. Please note that we ignore potential procedure in data compression and further refinement techniques through 2-D filtering in this case.

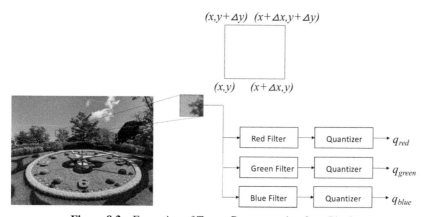

Figure 9.3 Formation of Tensor Representation for a Pixel.

Figure 9.4 Examples of partial mono-color object to be recognized (i.e. black car, yellow spiral, and red apple.

Figure 9.5 (Left) A lion on the grass land (Right) Partial edge information from the image.

9.1.2 Edge Detection

While studying the connection of brain and vision, it has been noted that certain neurons could be most excited by the edge at different orientation. Further psychology study demonstrated that only part of the image could be suffice to recognize the whole object as shown in Figure 9.4.

Actually, pixels described in Section 9.1.1 are not equally important to recognize from an image or a photo. It is noted that the edge information is particularly useful in many cases, which can be illustrated by Figure 9.5. A pretty complicated color image of a lion can be easily recognized by the edge information, even only partially available edges. Edges usually supply highly contracted information, which can be viewed as the high frequency components in signal analysis, to invoke stronger responses in the recognition of brain. Since the edges can be obtained by rather simple processing of an image, they are expected to useful in robot's perception of environments.

The purpose of *edge detection* that is extremely useful for vision is to identify sudden changes (i.e. discontinuities) in an image. Intuitively, most semantic and shape information from the image can be embedded into the edges of an object sine the edges assist extracting information, recognizing objects, and recovering geometry. In many cases, instead of thoroughly analyzing an image or a photo, the edges in an image can be very useful to

Figure 9.6 The edge information of a partial image of an object (i.e. car) is sufficient to recognize this object.

Figure 9.7 A photo taken from the parking lot in front of USF's Engineering Building.

detect and to recognize an object of interest. Together with the earlier lesson from psychology, Figure 9.6 depicts that even imprecise edge information of a partial image is enough to recognize an object.

▶ **Exercise:** Suppose you are designing the automatic parking functionality for an autonomous vehicle, with the following image from the camera on the top of the car (with height around 1.6 meter) as Figure 9.7, please develop a computer program to identify the parking spot. *Hint: Edge detection appears to work.*

▶ **Exercise:** In Figure 9.8, a photo taken in the garden of famous painter Monet, please apply edge detection to identify the house by noting some misleading curves. Most of the house is hidden behind plants but human brain can easily recognize it.

Figure 9.8 A photo taken in the Monet's garden.

9.1.3 Image Features and Object Recognition

Simply applying statistical decision theory by cross-correlation in Chapter 4 is actually not enough to achieve further scopes from images or video. Consequently, *local invariant image features* plays an important role in object detection, classification, tracking, motion estimation, etc. for image and video processing and understanding. The general methodology of employing local invariant features can be summarized as follows:

(a) Find and define a set of distinctive key-points.
(b) Define a local region around the key-point.
(c) Extract and normalize (or re-sizing) the regional content from the designated area.
(d) Compute a local descriptor from the normalized region (i.e., a function of pixel intensity or edges)
(e) Match local descriptors.

Due to the critical role of key-points, another technique known as *key-point localization* is developed, which detects features consistently and repeatedly, and allows for more precise localization to find interesting content within the image.

In the real world operation, a robot must recognize objects for proper actions, from different distances and angles, under different illuminations. Most importantly, the class of objects to be recognized might not have exactly the same appearance and thus image. For example, an autonomous vehicle must identify pedestrians for proper actions. Windows (e.g. in the Harris Corner detection) can be used to detect key-points. Using identically sized windows will not enable the detection of the same key-points across different-sized images. However, if the windows are appropriately scaled or re-sized, the same content or similar key-points can be captured. In addition to scaling in the window, the rotation of windows is another critical factor in this technique. The windowing techniques can be further enhanced by generalizing to 3D, from 8 neighboring pixels in 2D to 26 neighboring pixels in 3D.

In computer vision, it is of interest to identify groups of pixels that are observed together, which is known as *image segmentation*. Humans perform image segmentation by intuition. For instance, two persons looking at the same optical image might interpret in quite different manners, all depending on how their brains segment this image. Typical implementation of image segmentation is via clustering, which is introduced in the chapter of machine learning basics (e.g. K-means clustering, an unsupervised learning method).

By proper initialization, shifting, and rotating the window for the key-points, object recognition is to correctly assign the image to a cluster (i.e. a class of images to a correct label). If training is possible, K-nearest neighbors (KNN) is suitable as a supervised learning by appropriately selecting kernel functions. Object (and surely landmark) recognition is so important in localization and SLAM (in Chapter 7) for autonomous mobile robots.

▶ **Exercise:** Please apply above techniques to recognize the color objects (that is, which black object on the left appears highest similarity) in Figure 9.9. Ten black objects are clearly foot ball, shoe, horse, chicken, cat, tree, apple, microscope, house, and car.

Computer vision usually deals with visionary images (i.e. visible light). However, its principles can be generally applied to other imaging technologies, including radar techniques, in other frequency bands such as infrared, millimeter wave, etc. For the purpose of diverse information and human/data privacy, the methods other than traditional vision are often considered in robotics.

9.2 Multi-Modal Information Fusion Based on Visionary Functionalities

The most critical information to a robot such as an autonomous vehicle might be visionary sensors. For a human, the majority of brain operates to support

Figure 9.9 There are 10 black objects on the left and 5 color objects (in green, grey, red, yellow, and blue) to be recognized on the right.

vision. Due to the rich information contents from image or vision, visionary sensors supply critical information for robots. In this section, we pay primary attention to the visionary functionalities for autonomous vehicles. The typical visionary sensors include

- cameras operating at visible light to supply visionary information as images or videos
- radio detection and ranging (RADAR) using radio waveforms to detect subjects and their ranges, where millimeter wave radars get more attention due to line-of-sight propagation and purity in the frequency bands
- light detection and ranging (LIDAR) using visible or invisible light (say, infrared)
- ultrasonic sensors using pressure-based waves

Visionary sensors typically supply critical information for decisions by robots. Generally speaking, using autonomous driving as an example, knowledge representation of machine vision can be categorized to

- Metric knowledge, such as the geometry of static and dynamic objects, is required to keep the autonomous vehicle on the lane and maintain a safe distance to other machine/human-driving vehicles. Such knowledge typically includes multi-lane geometry, position and orientation of the own vehicle and the position or velocity of another traffic participant in the scenario.
- Symbolic knowledge, such as the classification of lanes as either "vehicle lane forward", "vehicle lane rearward", "bicycle lane", "walkway", or "intersection", allows to conform with basic operating rules.
- Conceptual knowledge, such as specifying a relationship between other traffic participants, allows to anticipate the expected evolution of the scene and to drive foresightedly, and to make appropriate decisions by the robot.

Example: The following Figure 9.10 illustrates different knowledges from an image for autonomous driving.

Figure 9.10 Cognitive Vision with metric knowledge (in yellow), symbolic knowledge (in orange), and conceptual knowledge (in red) [1].

Early vision prior to detection and classification of objects, supplies limited information from *single-frame* images, while *edge detection* typically proceeds based on the brightness or color gradients. In a contrast, *dual-frame* observation of a scene either from different visionary sensors or from a moving camera at different time yields images bearing significantly more information than single ones. The typical corresponding early vision tasks are *stereo vision* and *optical flow estimation*, respectively. Both methods fundamentally aim at finding precise and accurate matches between early-vision images, that is, pairs of corresponding pixel coordinates. An important characteristic is the density of results: Dense methods provide a match for ideally every pixel, and sparse methods usually only yield one match in a number of pixels. Since direct matching of raw images lacks efficiency and robustness, a *descriptor* that is computed from early vision ideally represents a pixel and its proximity in a way which allows for robust and efficient matching with other image. Finding the best match for a given image involves comparing and minimizing a dissimilarity measure, which is usually quantified by the Hamming distance for binary descriptors or the sum of absolute/squared differences for vector descriptors.

Stereo vision takes advantage of the fixed arrangement of predominantly multiple cameras which capture images at the same point in time: Rectification transforms these images such that the search space of matches can be limited to a one-dimensional disparity along a common image row, which helps to resolve ambiguities.

Optical flow represents the velocity and direction of image motion. The estimation of optical flow is a more general problem than stereo vision, because objects in the viewed scene as well as the camera itself may have moved arbitrarily between consecutive frames. Similarly to stereo rectification, their epipolar geometry can still be exploited for increasing a method's run- time efficiency, while in exchange limiting its scope to static scenes without moving objects. Such approaches enable 3-D reconstruction through structure-from-motion, and require the camera's ego-motion as an input.

To exploit further knowledge in 3D, *multi-frame* methods generally use more images than either a stereo or optical flow pair. For example, an important subset of such methods evaluates a quadruple of images from two cameras (stereo) at two points in time. This consequently enables the estimation of scene flow, a motion field representing the 3-D velocities of reconstructed 3-D points.

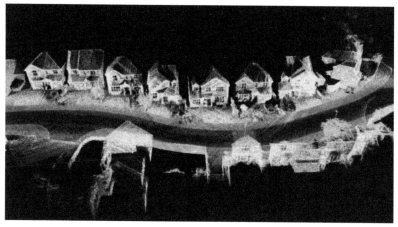

Figure 9.11 On-Board Capture of 3D LIDAR Information; http://www.meccanismocomple sso.org/en/laser-scanning-3d/.

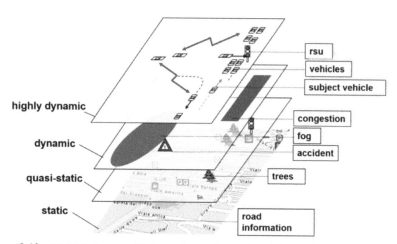

Figure 9.12 Multiple Layers of Information to Form a Local Dynamic Map; http://www.sa fespot-eu.org/documents.

To facilitate Multi-Modal Information Fusion, Figure 9.11 illustrates how to use on-board LIDAR forming 3D information, which is great for an autonomous vehicle to comprehend the driving environment.

However, to achieve the purpose of localization toward autonomous driving, multiple kinds of information must be fused. Figure 9.12 demonstrates one possible way to integrate such multi-modal information bu including map information and different levels of system dynamics.

In current (wireless) robotics, many different types of information may be considered in the multi-modal fusion, including data or information from

- on-board visionary devices: camera, LIDAR, mmWave radar, infrared sensor/imaging, etc.
- on-board localization devices that collect data to form desirable information: GPS, localization sensors, camera for landmarks, etc.
- on-board reference information: public/private (coordinate) references (i.e. map), pose estimation, gyro and odometer, belief of the robot state, etc.
- wireless networking: information that might influence planning (e.g. traffic situation and regulation), task assignments, states of other agents that potentially interact with the robot, predictive information of system operating, other sensor networks in the environment, etc.

Interesting readers may find more systematic methodology in [4].

9.3 Decision Trees

Making appropriate decisions always plays an important role in artificial intelligence. Please recall a serial of technologies that we have introduced. Hypothesis testing allows us to statistically make a decision based on the observation or a sequence of observations of the same. Sequential decision allows us to optimally decide from observations as time progresses. Markov decision process or reinforcement learning allows us to identify optimal strategy of decisions interacting with the environment. In this section, we will look into the methodology of making decisions based on rich structure of knowledge of information, which is known as *decision trees* and widely used in different aspects to realize the intelligence in robotics.

One of the most effective way to organize the structure of knowledge is through the tree structure that usually enjoys the computational advantages. Let us first review the fundamental property of tree, a special kind of graphs.

9.3.1 Illustration of Decisions

Example 12.1: A real-estate agent takes a good-taste customer Keiko, one newly married wife, to see a few houses on the market. This customer wants to find 3 candidate houses of interests to determine a final offer, but refuses to

Table 9.1 Features of various houses on the market, and the final interest level for this customer Keiko

BR No.	BA No.	Woods	Garden	Lot	Years	Open Kitchen	List Price	Interest Level
5	4	Yes	Big	Big	8	No	950K	No
2	2	Yes	Small	Big	5	Yes	450K	No
3	2	Yes	Small	Small	13	No	550K	No
4	2	No	Big	Small	2	Yes	650K	Yes
3	1	Yes	Big	Small	28	No	380K	No
4	3	No	Big	Small	23	Yes	690K	Yes
5	3	Yes	Big	Big	18	No	790K	No
4	3	No	Small	Small	1	Yes	580K	No
4	2	No	Big	Small	15	Yes	710K	?

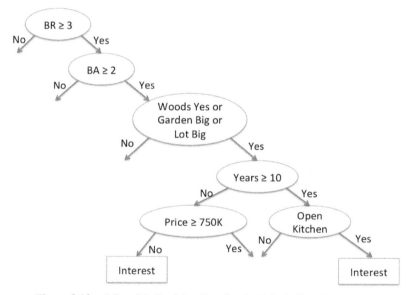

Figure 9.13 A Possible Decision Tree for the Keiko's Taste in Housing.

disclose factors to determine. The agent brilliantly makes a table (Table 12.1) to summarize the features of houses that Keiko has seen and the outcomes of Keiko's preference. How does the agent predict the interest level for the final candidate house?

This illustrated problem uniquely differs from the most common decisions to classify nominal and discrete data, without clear natural notation of similarity or even ordering. Patterns are described by a list of attributes rather than vectors of real numbers. The *decision tree* approach is particularly useful facing such non-metric data. We classify patterns through a sequence of questions, and the next question asked depends on the answer to the current question. Figure 12.1 shows a realization of decision tree corresponding to the situation in Table 12.1, and then the unknown interest level might be easily determined.

The remaining challenge is how to grow a tree based on the training samples and a set of features. A generic procedure known as *classification and regression tree* (CART) is described in the following as the straightforward illustration.

(a) We first identify tests, while each test or question involves a single feature of a subset of feature.
(b) A decision tree progressively splits the training set into smaller and smaller subsets.
(c) In case all samples at that node have the same class label, it is a pure node and thus no need to split further.
(d) Given data at a node, determine this node to be a leaf node or find another feature to split. This is known as recursive tree-growing process.

▶**Exercise:** Please construct the decision tree to separate red and green disks based on the Figure 9.14.

▶**Exercise:** Please construct the decision tree to separate squares of different colors based on the Figure 9.15.

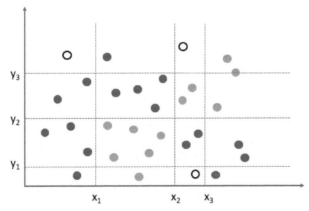

Figure 9.14   Disks in red and green distributed in two-dimensional plane, where black empty circles mean unable to determine color.

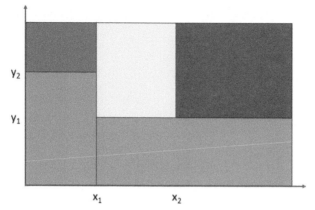

Figure 9.15 Disks in red and green distributed in two-dimensional plane, where black empty circles mean unable to determine color.

We are going to further details in the following.

9.3.2 Formal Treatment

In parametric estimation, a model is defined over the entire input space to learn associated parameters from the training data or observations. Then, we use the same model and the same set of parameters to infer from test input. For nonparametric estimation, the input space is divided into local regions according to specific distance measure (such as Euclidean distance), and the training data in this region gives the corresponding model.

Definition: A *decision tree* is a hierarchical model for supervised learning where a local region is recursively specified by a sequence of splits in a small number of steps. A decision tree consists of internal decision nodes and terminal leaves. Each *decision node* u implements a test function $f_u(\mathbf{x})$ with discrete outcomes labeling the branches. This process starts from the root and repeat recursively until a *leaf node*, whose value constitutes the output.

Remark: A decision tree is actually a nonparametric methodology since no parametric densities nor *a priori* knowledge of tree structure.

Remark: Each $f_u(\mathbf{x})$, where $\mathbf{x} = (x_1, x_2, \cdots, x_d)$, defines a discrimination function in the d-dimensional input space , to divide into small regions along the path from the root to leaves. $f_u(\cdot)$ is preferred a simple function such that a complex decision mechanism can be decomposed into simple decisions. A leaf node specifies a localized region in the input space while instances falling into this region have the same labels in classification or the same type of numeric outputs in regression.

Remark: A critical advantage of decision tree is the straightforward conversion to a set of if-then logic rules that is easy in implementation.

Definition: In a *univariate tree*, the test of each internal decision node uses only one of the input dimensions.

In other words, the test for node u can be represented as simple as a binary decision

$$f_u(\mathbf{u}) : x_j \geq \eta_{uj} \tag{9.1}$$

or

$$f_u(\mathbf{u}) : x_j < \eta_{uj} \tag{9.2}$$

where η_{uj} is the selected decision threshold. The decision tree in Example 12.2 is univariate.

9.3.3 Classification Trees

If a decision tree is established for the purpose of classification (or hypothesis testing), it is named as *classification tree*, in which the goodness of a split is quantified as the *impurity measure*. A split is *pure* if all the instances in any branch belong to the same class after the split.

An intuitive way to measure the impurity is to estimate the probability of class C_i. For node u, suppose N_u denotes the number of training instances reaching node u, where N for the root node. N_u^i of N_u belong to class C_i, and $\sum_i N_u^i = N_u$. Given an instance reaching node u, the estimate of probability of class C_i is intuitively obtained as

$$\hat{p}_u^i = P(C_i|\mathbf{x}, u) = \frac{N_u^i}{N_u} \tag{9.3}$$

Node u is pure if p_u^i are either 0 or 1 for all i. The value 0 means none of the instances reaching node u are in class C_i, and the value 1 means all such instances in C_i. If the split is pure, we do not split further and can amend a leaf node labeled with the class. A popular measure of impurity for classes $\{C_1, C_2, \cdots, C_K\}$ uses entropy as

$$I_u = -\sum_{i=1}^{K} p_u^i \log p_u^i \tag{9.4}$$

As a matter of fact, possible measure for such two-class problem can be realized as a nonnegative function ϕ satisfying

- $\phi(p, 1-p), p \in [0, 1]$, reaching maximum when $p = 1/2$.
- $\phi(0, 1) = \phi(1, 0) = 0$
- $\phi(p, 1-p)$ is increasing on $[0, 1/2]$ and decreasing on $[1/2, 1]$

A few examples satisfy above conditions for $\phi(p, 1-p)$ and can well serve the purpose of impurity measure:

- Binary entropy function: $\phi(p, 1-p) = h_b(p)$
- Gini index: $\phi(p, 1-p) = 2p(1-p)$
- Misclassification error: $\phi(p, 1-p) = 1 - \max(p, 1-p)$

All of these measures can be extended into multiple (i.e. ≥ 2) classes and no significance among these three measures.

If the node u is not pure, it is desirable to find the split that minimizes impurity after the split in order to generate the smallest tree. Suppose at node u, N_{uj} of N_u take branch j and there exist $\mathbf{x}^t$ such that the test $f_u(\mathbf{x}^t)$ returns outcome j. The estimate for probability of class C_i is

$$\hat{P}(C_i | \mathbf{x}, u, j) = \frac{N_{uj}^i}{N_{uj}} \tag{9.5}$$

where $N_{[uj]}^i$ of N_{uj} belong to class C_i and $\sum_{i=1}^{K} N_{uj}^i = N_{uj}$. The subsequent total impurity after the split is

$$I_u^* = -\sum_{j=1}^{n} \frac{N_{uj}}{N_u} \sum_{i=1}^{K} p_{uj}^i \log p_{uj}^i \tag{9.6}$$

where $\sum_{j=1}^{n} N_{uj} = N_u$

▶ **Exercise:** Please summarize the algorithm to construct a classification tree.

Remark: Above principle sets the foundation of *classification and regression tree* (CART) algorithm and its extensions. At each step of tree construction, we select the split causing the largest decrease in impurity, which is the difference between two equations.

9.3.4 Regression Trees

A regression tree is constructed pretty much the same as a classification tree, except considering appropriate impurity measure for regression. For node u,

X_u is the subset of X reaching node u. In other words, it is the set of all $\mathbf{x} \in X$ satisfying all conditions in the decision nodes on the path from the root to node u. We define a function to indicate $\mathbf{x}$ reaching node u as

$$b_u(\mathbf{x}) = \begin{cases} 1, & \mathbf{x} \in X_u \\ 0, & \text{otherwise} \end{cases} \tag{9.7}$$

In regression, the goodness of a split can be measured by the mean squared error (MSE) from the estimated value. Let g_u denote the estimated value in node u.

$$V_u = \frac{1}{|X_u|} \sum_t (r^t - g_u)^2 b_u(\mathbf{x}^t) \tag{9.8}$$

where $|X_u| = \sum_t b_u(\mathbf{x}^t)$. In a specific node, the mean of the required outputs of instances reaching the node is used to define

$$g_u = \frac{\sum_t b_u(\mathbf{x}^t) r^t}{\sum_t b_u(\mathbf{x}^t)} \tag{9.9}$$

Therefore, V_u corresponds to the variance at node u. If the error is acceptable, that is, $V_u < \eta_r$, then a leaf node is created and stores the value of g_u. If the error is not acceptable, data reaching node u is split further such that the sum of errors in the branches is minimum. The process continues recursively, similar to decision tree. We can define X_{uj} as the subset of X_u taking branch j and $\cup_{j=1}^n X_{uj} = X_u$. In a similar way, $b_{uj}(\mathbf{x})$ denotes the indicator of $\mathbf{x} \in X_{uj}$ reaching node u and taking branch j and g_{uj} denotes the estimated value in branch j at the node u.

$$g_{uj} = \frac{\sum_t b_{uj}(\mathbf{x}^t) r^t}{\sum_t b_{uj}(\mathbf{x}^t)} \tag{9.10}$$

Then, the error after the split is

$$V_u^* = \frac{1}{|X_u|} \sum_j \sum_t (r^t - g_{uj})^2 b_{uj}(\mathbf{x}^t) \tag{9.11}$$

Remark: Instead of taking average at a leaf that implements a constant fit, a linear regression fit over instances can be adopted.

$$g_u(\mathbf{x}) = \mathbf{w_u^T} \mathbf{x} + w_{u0} \tag{9.12}$$

9.3.5 Rules and Trees

We have introduced the methods to grow the trees by successive splitting nodes. However, how can we terminate? We can continue until each terminal node contains only one instance, which leads to the overfitting problem (i.e. great performance in training but poor performance in predictions. Two basic approaches are applicable:

- Stopping Rule: Stop splitting the node if the change in the impurity function is less than a pre-determined threshold, which is hard to specify good threshold in advance.
- Pruning: A decision tree grows until its terminal nodes have pure class, and then prune it, replacing a subtree with a terminal node.

To make inference over a decision tree computationally feasible, rules can be developed to deduct the possible nodes to visit/compute. The following exercise demonstrates this application.

■ **Exercise:** Suppose blue circle and green circle representing police cars, which can observe the streets (in white) vertically and horizontally until the blocking (in black), that is, line-of-sight. Without observing theft car (in red star), a police car moves randomly, that is, equally probable to move forward or turn left and right. Once a police car observes any theft car, it will keep tracing until the theft car being blocked by two police cars. During the tracing, the policy might lose line-of-sight observation to determine its movement in a random manner if no inference result is available. All the cars know the street map as Figure 9.16, while the blue police car can observe the red-star theft car, and notify green police car about the position of the red-star theft car. The red-star theft car now move one step first, then police cars move one step alternatively, while all cars can move one step at a time following the decision tree it builds. The depth of each decision tree can not go beyond 64 levels. Can red-star car run away to one of three exists before being blocked by two police cars?

9.3.6 Localizing A Robot

Decision tree is a useful method to enhance intelligence in many robotic applications. For example, as shown in Figure 9.17, the robot knows the map polygon $\mathcal{P}$ and the visibility polygon $mathcalV$ representing what the robot can see in the environment from its present location. Suppose also that the robot knows that $\mathcal{P}$ and $\mathcal{V}$ should be oriented as shown. The black dot represents the robot's position in the visibility polygon. By examining $\mathcal{P}$ and

Figure 9.16   Street Map (black: prohibited driving; white: street).

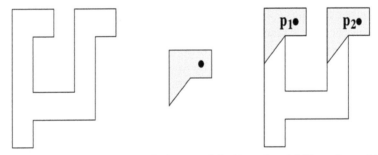

Figure 9.17   Given a map polygon $\mathcal{P}$ shown at left side and a visibility polygon $\mathcal{V}$ in the middle, the robot must determine which of the two possible initial locations p_1 and p_2 at the right side is its actual location in $\mathcal{P}$ [5].

$\mathcal{V}$, the robot can determine that it is at either point p_1 or point p_2 in $\mathcal{P}$, which implies the set of hypotheses to be $\mathcal{H} = \{p_1, p_2\}$.

It is not distinguishable between p_1 and p_2 since $\mathcal{V}(p_1) = \mathcal{V}(p_2) = \mathcal{V}$. However, by traveling to the hallway section in $\mathcal{P}$ and taking another probe, the robot is able to precisely determine its location. Such a *verification tour* is a tour along which a robot that knows its initial position a priori can travel to verify this information by probing and then return to its starting position. An optimal verification tour is a verification tour of minimum traveling length d_{min}.

There is no need to assume any *a priori* knowledge or statistics of $\mathcal{H}$. Decision tree can be therefore applied. A localizing decision tree is a tree consisting of two kinds of nodes and two kinds of weighted edges. The nodes are either sensing nodes (S-nodes) or reducing nodes (R-nodes), and the node types alternate along any path from the root to a leaf. Thus, tree edges directed down the tree either join an S-node to an R-node (SR-edges) or join an R-node to an S-node (RS-edges).

1. A localizing decision tree is a tree consisting of two kinds of nodes and two kinds of weighted edges. The nodes are either sensing nodes (S-nodes) or reducing nodes (R-nodes), and the node types alternate along any path from the root to a leaf. Thus, tree edges directed down the tree either join an S-node to an R-node (SR-edges) or join an R-node to an S-node (RS-edges).

2. Each R-node is associated with a set $\mathcal{H}' \subseteq \mathcal{H}$ of hypothetical initial locations that have not yet been ruled out. The root is an R-node associated with $\mathcal{H}$, and each leaf is an R-node associated with a singleton hypothesis set.

3. Each SR-edge represents the computation that the robot does to rule out hypotheses in light of the information gathered at the S-node end of the edge. An SR-edge does not represent physical travel by the robot and hence has weight 0.

4. Each RS-edge has an associated path defined relative to the initial location of the robot. This is the path along which the robot is directed to travel to reach its next sensing point. The weight of an RS-edge is the length of its associated path.

In order to minimize the traveling distance of the robot, the weighted height of a localizing decision tree can be defined in the following manner. The weight of a root-to-leaf path in a localizing decision tree is the sum of the weights on the edges in the path. The weighted height of a localizing decision tree is the weight of a maximum-weight root-to-leaf path. An optimal localizing decision tree is a localizing decision tree of minimum weighted height. Generally speaking, the problem of finding an optimal localizing decision tree is NP-hard.

▶ **Exercise:** Please develop the localizing decision tree for Figure 9.18 with $\mathcal{H} = \{p_1, p_2, p_3, p_4\}$. Please properly select the probing point(s) that is related to the height of localizing decision tree. Please find the complexity to execute the search over decision tree in order to obtain minimum traveling distance for this robot.

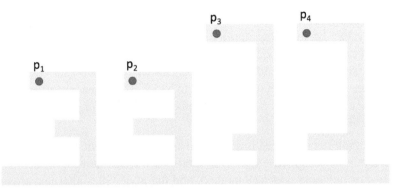

Figure 9.18   Localizing a robot among four possible locations, given the viewing angle of this robot to be 90°.

9.3.7 Reinforcement Learning with Decision Trees

In many robotic or agent-based applications, acquiring experiences can be very expensive and time-consuming. Two of the main approaches towards this goal are to incorporate generalization (function approximation) into model-free methods and to develop model-based algorithms. Model- based methods achieve high sample efficiency by learning a model of the domain and simulating experiences in their model, thus saving precious samples in the real world. Once a model-based reinforcement learning method has built an accurate model of the domain, it can quickly find an optimal policy by performing value iteration within its model. Thus the key to making model-based methods more sample efficient is to make their model learning more efficient. One way to learn the model of the domain quickly is to introduce generalization into the learning of the model using modern machine learning techniques. Learning the model is essentially a supervised learning problem where the input is the agent's current state and action, and the learning algorithm has to predict the agent's next state and reward. Many existing supervised learning algorithms are able to generalize their predictions to new or unseen parts of the state space.

Learning the model efficiently requires a combination of a fast learning algorithm and a policy that acquires the necessary training examples quickly. The agent can target states for exploration that it expects will improve the model. These states could be where the agent has not visited frequently or where the model has low confidence in its predictions. To achieve above wish, a novel reinforcement learning algorithm called *Reinforcement Learning with Decision Trees* (RL-DT) will be introduced, which uses decision trees to learn a model of the domain efficiently, incorporating generalization into the learning of the model. The algorithm explores early to learn an accurate model before switching to an exploitation mode.

The mathematical formulation of RL-DT starts from standard MDP as Chapter 4, which consists of a set of states $\mathcal{S}$, a set of actions $\mathcal{A}$, a reward function $R(s, a)$, and a state-transition function $P(s' \mid s, a)$. In a state $s \in \mathcal{S}$, the agent takes an action $a \in \mathcal{A}$, then receives a reward $R(s, a)$, to reach a new state s' by $P(s' \mid s, a)$. The goal of the agent is to find a policy π mapping states to actions that maximizes the expected total reward discounted over the horizon. The value $Q^*(s, a)$ of any given state-action pair is determined by solving the Bellman equation:

$$Q^*(s, a) = R(s, a) + \gamma \sum_{s'} P(s' \mid s, a) \max_{a'} Q^*(s', a') \tag{9.13}$$

where $0 < \gamma, 1$ denotes the discount factor. The optimal value function Q^* can be computed through the value iteration over the Bellman equations toward convergence. The optimal policy π is

$$\pi(s) = \operatorname*{argmax}_a Q^*(s, a) \tag{9.14}$$

The Rl-DT algorithm is a model-based reinforcement learning algorithm, which maintains the set of all the states in the observed set $\mathcal{S}_M$ and counts the number of visits to each state-action pair. From a given state s, a robot executes the action a as specified by its action-values, and increments the visit count $Visit(s, a)$. It obtains a reward r and a next state s. It adds the state to its state set $\mathcal{S}_M$ if it has not been already there. Then the algorithm updates its model with this new experience through the *model learning* approach described later. The algorithm decides to explore or exploit based on whether it believes its model is accurate, which relies on the *Check-Model* described later. If the model is changed, the algorithm will re-compute action values by value iterations.

A special purpose of RL-DT is to learns a model of the transition and reward functions, particularly to learn a model of the underlying MDP in as few as possible samples. It is noted that the transition and reward functions in many states may be similar, and if so, the model can be learned much faster by generalizing these functions across similar states, which is particularly true to learn an accurate model of the MDP without visiting every state. For example, in many grid examples in this book, it is easier to generalize the relative effect of moving right (say, $x \leftarrow x + 1$) than absolute effects (say,

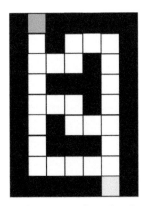

Figure 9.19   Start from green tile and end at the yellow tile.

$x \leftarrow 5$ or $x \leftarrow 6$). RL-DT takes advantage of this rule by using supervised learning techniques to generalize the relative effects of actions across states when learning its model, which can predict the effects of actions even for those states that it has not visited often or at all.

The agent learns models of the transition and reward functions using decision trees. A separate decision tree is built to predict the reward and each of the n state variables. The first n trees each make a prediction of the probabilities $P(x_i^r \mid s, a)$, while the last tree predicts the reward $R(s, a)$. Once the model has been updated, the value iteration is performed on the model to find the desired policy.

We use the following example to illustrate how to apply decision tree into earlier mobile robot scenario.

Example: To walk through the maze as shown in Figure 9.19, the pre-designated decision mechanism may (i) prefer straight (ii) prefer left turn (iii) prefer right turn (iv) randomize direction to proceed at the interaction when walking through the room. How can we establish decision tree(s) to rapidly learn the preference among four alternatives for this predesignated mechanism?

An agent has own preference for moving pattern; walk straight, turn left, and turn right. To mimic one agent moving pattern, the observing agent need to observe the target agent's moving pattern first. An agent walk through the block showed in Figure 9.19. The structure is that the green block is the starting point and the agent wants to reach to the goal (i.e. yellow block). This trial will be executed N times, then after N trials, the observing agent generates the observed agent's moving model to guess the moving preference.

Algorithm for mimicking the movement

First, to keep the example simple, the observing agent assumes the target agent moves based on Markov Decision Process (MDP).The procedure for mimicking is as below.

- Creating the tree of states (position)
- Observing N trials of the agent's moving pattern
- Simulating the path using the tree which was generated in the first procedure

Creating the tree

It is assumed that the observing agent knows the whole structure of the map and all possible blocks to pass from Figure 9.19. Here, the observing agent creates tree of each state s (position of block). And the agent also knows the all neighbor states of s, moreover, possible neighbor states to pass for next step. Now we define the list of the possible neighbors (for example $s', s'', \ldots$,) of state s as $neighbor(s) = [s', s'', \ldots]$. Then the agent connect all states in the list $neighbor(s)$ to state s, and iterate this procedure to all states. In this tree, state $s', s'', \ldots$ in $neighbor(s)$ can be called as $child$ of state s, and also state s can be regarded as $parent$ of $s', s'', \ldots,$.

- Begin with the start state $s^{(s)}$ and get $neighbor(s^{(s)})$
- Connect the all the neighbors in $neighbor(s^{(s)})$ to state $s^{(s)}$ with edges
- Go to the next state in $neighbor(s^{(s)})$, and apply the same procedure for each state
- Iterate until reaching the goal state $s^{(g)}$

Observing and simulating the tree

Now, the observing agent simulates the tree based on the target agent's trials. What the observing agent needs to do is to simulate the whole path to the state nodes on the tree. During N trials of the target, the agent counts the number of visiting state s as N_s for each node (state). When state s is a parent and the s' is a child, we can define the probability of the transition from state s to s' as $p_{ss'} = p(s'|s)$. This $p_{ss'}$ suggests that the agent's preference to go to state s' when the agent at the state s. If s' is the only state in the list $neighbor(s)$, in other words, is the only one child of s, then the $p_{ss'} = 1$, which means the agent just walks straight from state s to s'. However, if there exist multiple child nodes from state s, $p_{ss'} = N_{s'}/N_s$ while N_s' is the number of visiting the state of the child nodes s'. The procedure of simulating the tree is as below

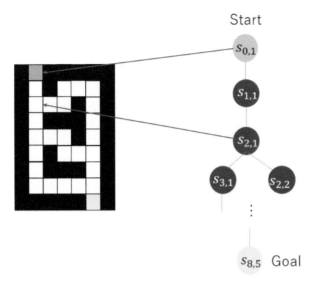

Figure 9.20　Creating the tree.

- Initial setting, $n = 0$, n is the number of trials executed, and $N_s = 0$ for $s \in S$, S is the set of the all states
- $n \leftarrow n + 1$
- At node s, if the target visited the state s, $N_s \leftarrow N_s + 1$ and applying same procedure to all the state $s \in S$
- Repeat the same procedure from the second until $n = N$
- Check the $p_{ss'}$ if the node of the state s and s' is connected and s is the parent of the state s', then calculate $p_{ss'} = N_{s'}/N_s$

Simulations

Using the Figure 9.19 and simulate the target agent moving pattern with $N = 10$, and the observing agent tries to get the preference of the target.

　　Figure 9.20 shows how to generate the tree based on the map in Figure 9.19. In the figure, the starting state is $s = s_{0,1}$, then the agent generates the tree until reaching the goal $s_{8,5}$. After generating the tree and when the agent simulates the tree, Figure 9.21 shows how to go to the each node of state based on Figure 9.19. The example is when the agent walk through the block $(0, 1), (1, 1), (2, 1), (3, 1), \ldots$, then on the tree, the agent simulate through $s_{0,1}, s_{1,1}, s_{2,1}, s_{3,1}, \ldots$.

　　After N trials, the agent calculate $p_{ss'}$ for each edge. In the example of Figure 9.22, there are two intersection points that means those state has 2

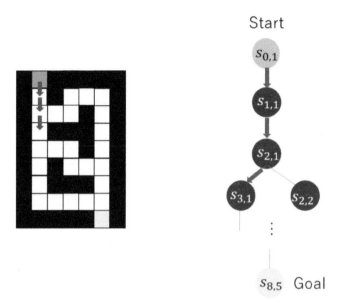

Figure 9.21   Simulating the tree.

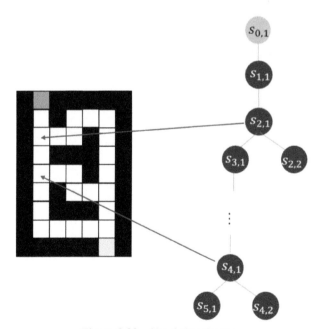

Figure 9.22   Simulating the tree.

child nodes in the tree. In this example, the agent calculate $p_{ss'} = p(s'|s)$ and the result was $p(s_{3,1}|s_{2,1}) = 0.67, p(s_{2,2}|s_{2,1}) = 0.33, p(s_{5,1}|s_{4,1}) = 0.67$, and $p(s_{4,2}|s_{5,1}) = 0.33$; with other states s and s', $p(s'|s) = 1$. In this case, observing agent concludes that the target agent's preference is going to straight with probability $p = 0.67$ and turning right with $p = 0.33$ when there is an intersection.

We use this simple example to illustrate how to take advantage of decision tree to learn another agent's behavior model, which has broad-range applications.

Remark: RL-DT presents a variant of RL to rapidly learn the model of MDP, which can be applied in many scenarios, namely man-robot collaboration, robot soccer games, and chasing vehicle in earlier exercise of this chapter.

9.4 Federated Learning

Fusion center to conduct inference of collected data through wireless links is a common technological scenario to facilitate AI in IoT, which indicates a technical challenge to preserve security or at least privacy of the transmitted data, also to the application scenarios of 6G mobile communications. When devices can have good computing, ML such as deep learning serves the means of inference, while the privacy preserving of involved data (or datasets) is considered high priority technical goal to accomplish. Recently, a new methodology of ML, known as *federated learning* (FL), is proposed to resolve this technical challenge [7] and will be introduced in this section.

9.4.1 Federated Learning Basics

Suppose there are N data owner (or sensors) $\{S_i\}_{i=1}^N$ in a network, and S_i has a dataset $\mathbb{D}_i$. Traditionally, the fusion center must collect all datasets $\{\mathbb{D}_i\}_{i=1}^N$ through wireless channels to conduct inference by training the ML model, M_{SUM}, which implies an undesirable fact that S_i has to open $\mathbb{D}_i$ to fusion center.

On the contrary, federated learning (FL) aims to build a joint ML model based on these datasets, without actual obtaining any dataset. FL proceeds in two stages, *model training* and *model inference*. In model training, information can be exchanged. but not the data, without revealing any private portion of $\{\mathbb{D}_i\}_{i=1}^N$. The trained model can reside at fusion center or any S_i, or collaboratively shared among them. The model can be applied to inference on any new dataset:

- $\{S_i\}_{i=1}^{N}$ jointly build a ML model, while each of them collaboratively contribute some data/information to train the model.
- In the stage of model training, the data held by each data owner is not actually disclosed.
- The mode can be transform among the data owners (and fusion center) in a secure communication manner (.eg. in encryption) such that the reverse-engineering of data is not possible.
- The resulting model, M_{FED} is a good approximation to the ideal model by direct using all the datasets.

Denote L_{SUM} and L_{FED} as the performance measures of M_{SUM} and M_{FED}. Let $\delta > 0$, M_{FED} achieves δ-performance loss if

$$| L_{FED} - L_{SUM} | < \delta \qquad (9.15)$$

The dataset $\mathbb{D}_i$ can be viewed as a matrix, where each row represents a (data) sample and each column indicates a specific feature, with possible label for a sample. The feature space and label space can be denoted as $\mathcal{X}$ and $\mathcal{Y}$ respectively, and the sample ID space is denoted as $\mathcal{I}$. The label might indicate the annual charge of policy in an insurance dataset, or the purchasing desire in an e-commerce dataset.

Figure 9.23 indicates two basic forms of FL. *Horizontal federated learning* (HFL) is also known as sample-partitioned FL or example-partitioned FL. HFL is suitable to infer while datasets at different sites share overlapping feature space but differs in sample space. The conditions for HFL are

$$\mathcal{X}_i = \mathcal{X}_j, \mathcal{Y}_i = \mathcal{Y}_j, \mathcal{I}_i \neq \mathcal{I}_j, \forall \mathbb{D}_i, \mathbb{D}_j, i \neq j \qquad (9.16)$$

On the other hand, If the identity and the status of each participant is the same, the federation establishes a "commonwealth" strategy, and is known as

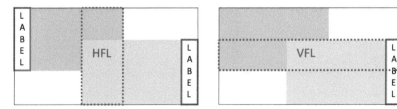

Figure 9.23 (left) Horizontal Federated Learning (right) Vertical Federated Learning [7].

vertical federated learning (VFL) with conditions

$$\mathcal{X}_i \neq \mathcal{X}_j, \mathcal{Y}_i \neq \mathcal{Y}_j, \mathcal{I}_i = \mathcal{I}_j, \forall \mathbb{D}_i, \mathbb{D}_j, i \neq j \qquad (9.17)$$

Wireless (sensor) data collection and fusion/inference while privacy is a serious concern (say, privacy-preserving data collection) appears to be a scenario feasible for FL by treating each data source generating a data set. Both HFL and VFL are possible depending on exact operating conditions.

9.4.2 Federated Learning Through Wireless Communications

It is usually interesting to consider federated learning (FL) with deep learning (DL). Figure 9.24 illustrates an application scenario, under the desirable privacy preserving requirement, each fusion center is capable of deep learning to infer from own sensor data, and a centralized DL mechanism can communicate with these fusion centers toward deep learning on all sensor data. For privacy/security or bandwidth concerns, particularly the communication links between the centralized DL mechanism and fusion centers are wireless, it might not be appropriate to send the raw datasets from fusion centers to centralized DL mechanism/server.

FL reveals an innovative procedure to achieve the purpose of deriving the DL model from the whole data but without transmitting the data as follows.

1. The centralized DL trains the global model (i.e. deep learning model).
2. This global model is sent to each distributed fusion center.
3. Each fusion center optimize the model (i.e. fusion model) by using own dataset collected from its sensor network.

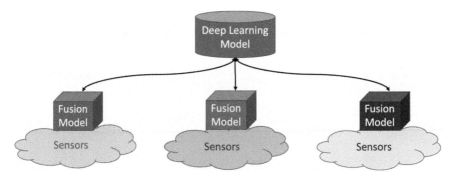

Figure 9.24   A centralized deep learning to communicate with several sensor fusion centers corresponding to sensor networks.

4. These locally trained models are uploaded to the centralized DL mechanism/server for updates.
5. Apply an appropriate FL model (say, FedAvg or FedSGD) to obtain the new global DL model.
6. If convergence criterion is satisfied, FL is accomplished. Otherwise, repeat the same procedure as above.

The FedSGD algorithm proceeds in an intuitive way since DL typically proceeds by applying the gradients to train neural networks. Suppose, among N fusion centers, each fusion center conducts DL in a smaller dataset to get its fusion model with gradient g_n. The centralized DL server computes its gradient

$$g = \frac{1}{N} \sum_{n=1}^{N} g_n \tag{9.18}$$

The weights in the NN can be obtained by

$$\mathbf{w}_{new} = \mathbf{w}_{old} - \gamma \cdot g \tag{9.19}$$

where γ denotes the learning rate. To avoid frequent communication between the centralized server and distributed fusion centers, an effective and widely applied algorithm, FedAvg, is developed by simply averaging the weights

$$\mathbf{w}_{new} = \frac{1}{N} \sum_{n=1}^{N} \mathbf{w}_n \tag{9.20}$$

Above description of FL simply assumes the existence of communication links. State-of-the-art research exploits new methods for FL and explores further by taking wireless communications into consideration.

9.4.3 Federated Learning over Wireless Networks

We initially consider the problem of FL over wireless networks as Figure 9.24 over a multi-user wireless system consisting of a base station (BS) connected to the centralized DL model and N UEs serving these N fusion models [8]. Each participating UE n stores a local dataset $\mathbb{D}_n$ of size D_n. The subsequent total data size is

$$D = \sum_{n=1}^{N} D_n \tag{9.21}$$

Considering the supervised learning, at UE n, $\mathbb{D}_n$ represents the collection of data given a set of input-output vector pairs $\{x_i, y_i\}_{i=1}^{D_n}$. For such a supervised learning problem, the goal is to find the model parameter w that characterizes the output y_i with loss function $f_i(w)$. For example, a common form is

$$f_i(w) = \frac{1}{2}\|x_i^T w - y_i\|^2 \tag{9.22}$$

The loss function on the dataset $\mathbb{D}_n$ at the UE n is therefore

$$J_n(w) = \frac{1}{D_n} \sum_{i \in \mathbb{D}_n} f_i(w) \tag{9.23}$$

The learning model minimizes the following global loss function

$$\min_w J(w) = \sum_{n=1}^{N} \frac{D_n}{D} J_n(w) \tag{9.24}$$

The FL proceeds [9] as

1. For each UE, at the tth update,
   Computation: Each UE solves its local problem

$$w_n^t = \operatorname{argmin}_{w_n} F_n(w_n \mid w^{t-1}, \nabla J^{t-1}) \tag{9.25}$$

with a local accuracy $0 \leq \theta \leq 1$.

Communication: Each UE sends w_n^t and ∇J_n^t to BS, according to a pre-designated networking mechanism (say, synchronous time-slotted reservation known as TDMA).

2. At the BS, the following information is aggregated

$$w^{t+1} = \frac{1}{N} \sum_{n=1}^{N} w_n^t \tag{9.26}$$

$$\nabla J^{t+1} = \frac{1}{N} \sum_{n=1}^{N} \nabla_n^t \tag{9.27}$$

Then, feedback to all UEs. This process repeats until reaching the global accuracy $0 \leq \epsilon \leq 1$ (i.e. $\|\nabla J(w^t)\|$ converges).

With this setting, FL over wireless networking problem faces the trade-offs of (i) learning time versus UE energy consumption by using Pareto efficiency model, and (ii) computation versus communication learning time by finding the optimal learning accuracy parameter [8]. To understand these questions, we adopt the computation model and the communication model as follows.

- For each UE n, computing $\mathbb{D}_n$ takes $c_n D_n$ instruction cycles. With CPU-cycle frequency ϕ_n at UE n, the CPU energy consumption of UE n for one local iteration of computation can be expressed as

$$\mathcal{E}_n^{comp}(\phi_n) = \sum_{i=1}^{c_n D_n} \frac{\alpha_n}{2}\phi_n^2 = \frac{\alpha_n}{2}c_n D_n \phi_n^2 \qquad (9.28)$$

  where $\alpha_n/2$ is the capacitance coefficient of CPU energy efficiency. The computing time is thus $c_n D_n/\phi_n$.
- Assuming OFDM over AWGN channel, the achievable transmission rate of UE n is

$$r_n = B\log(1 + \frac{h_n P_n}{N_0}) \qquad (9.29)$$

  where B denotes the bandwidth; h_n denotes the channel gain constant; P_n represents the transmission power; N_0 denotes the additive noise.

Ideally, the data size of both w_n and ∇J_n is s_n, which takes the fraction τ_n of r_n to transmit. That is,

$$\tau_n = s_n/r_n \qquad (9.30)$$

as the most energy-efficient transmission policy, which give the communication efficiency $\mathcal{E}_n^{comm}(\tau_n)$. Consequently, we can form the corresponding optimization problem to explore the trade-offs.

9.4.4 Federated Learning over Multiple Access Communications

In earlier modeling, we assume an ideal networking mechanism in the Fl over wireless networking or precisely multiple access communications. Practically, we either apply a multiple access protocol or adopt a reservation mechanism with control signaling that consumes extra bandwidth. For a specific learning/fusion model, deep learning or linear regression, the loss function is defined as (9.22) and the w_n^{t+1} is locally updated by x_n, y_n, w^t. BS updates the DL (or regression) model as (9.26).

In FL, local updating and aggregation at the BS is iteratively carried out by (9.22) and (9.26). This iteration requires uploading the local weight vectors from N UEs. If N is large, the required communication time for uploading per iteration might be long in this multiple access communications. To shorten the uploading time, multiple channels can be used with the multiple access protocol. Considering the simplest multiaccess protocol ALOHA, we intend to develop a random sampling approach to approximate for (9.26) via multi-channel ALOHA.

For original FL in earlier subsection, the averaging in (9.26) requires all the N local updates. However, wireless communications always needs to deal with a critical limitation, bandwidth. We have to evaluate the feasibility that all the local updates may not be available at each iteration. In other words, suppose that there are M parallel channels, where $M \ll N$ so that M UEs can simultaneously upload their local updates at each iteration. To avoid the selection dilemma, multi-channel ALOHA with the access probability that depends on the local update is employed.

We first formulate an optimization problem to approximate the aggregation in terms of the access probabilities of UEs based on the BS?s perspective. Then, we show that each user can decide its access probability with its local update and (simple) feedback information from the BS. Define $a = \sum_{n=1}^{N} w_n$, which is the unnormalized aggregation. Further define $u = \sum_{n=1}^{N} w_n \delta_n$, $\delta_n \in \{0, 1\}$. , where $\delta_n = 1$ serves the indicator if the BS receives the local update from UE n. We assume that δ_n is dependent on w_n. u can be viewed as an approximation of a for the aggregation in (9.26). To evalluate the approximation error, we can consider the following conditional error norm:

$$\mathbb{E}\left[\|a - u\| \mid \mathcal{W}\right] = \mathbb{E}\left[\|\sum_{n=1}^{N} w_n(1 - \delta_n)\| \mid \mathcal{W}\right] \tag{9.31}$$

$$\leq \sum_{n=1}^{N} a_n \mathbb{E}\left[1 - \delta_n \mid w_n\right] \tag{9.32}$$

$$\leq \sum_{n=1}^{N} a_n e^{-q_n} \tag{9.33}$$

where $a_n = \|w_n\|$ and $q_n = \mathbb{E}\left[\delta_n \mid w_n\right]$ is the probability that BS receives the local update from UE n. (9.32) comes from the triangle property and (9.33) is due to the fact that $1 - x \leq e^{-x}, x \in (0, 1)$.

[10] evaluates FL over multi-channel ALOHA based on above principle to demonstrate decent performance. [11, 12] also present interesting explorations on the nature of wireless communications.

When robots have to rely on wireless sensor networks to execute precise and safe operations, federated learning and other privacy-preserving inference techniques emerge a critical aspect of wireless robotics.

Further Reading: [2] has detailed materials regarding computer vision and imaging for autonomous vehicles. [3] supplies more detailed knowledge and techniques about computer vision. [4] supplies a great overview on multimodal data fusion. [6] presents an early effort integrating decision tree and RL.

References

[1] B. Ranft, C. Stiller, "The Role of Machine Vision for Intelligent Vehicles", *IEEE Tr. on Intelligent Vehicles*, vol. 1, no. 1, pp. 8–19, March 2016.

[2] R.P. Loce, R. Bala, M. Trivedi, *Computer Vision and Imaging in Intelligent Transportation Systems*, Wiley-IEEE, 2017.

[3] R. Szeliski, *Computer Vision: Algorithms and Applications*, Springer, 2010.

[4] D. Lahat, T. Adali, C. Jutten, "Multimodal Data Fusion: An Overview of Methods, Challenges, and Prospects", *Proceeding of the IEEE*, vol. 103, no. 9, pp. 1449–1477, Sep. 2015.

[5] G. Dudek, K. Romanik, S. Whitesides, "Localizing A Robot With Minimal Travel", *SIAM J. Comput.*, vol. 27, n0. 2, pp. 583-604, April 1998.

[6] T. Hester, P. Stone, "Generalized Model Learning for Reinforcement Learning in Factored Domains", *The Eighth International Conference on Autonomous Agents and Multiagent Systems (AAMAS 09)*, Budapest, 2009.

[7] Q. Yang, Y. Liu, T. Chen, and Y. Tong, "Federated machine learning: Concept and applications", *ACM Trans. Intell. Syst. Technol.*, vol. 10, pp. 12:1–19, Jan. 2019.

[8] N.H. Tran, W. Bao, A. Zomaya, M.N.H. Nguyen, C.S. Hong, "Federated Learning over Wireless Networks: Optimization Model Design and Analysis", *IEEE INFOCOM*, 2019.

 [9] J. Konecny, H. B. McMahan, D. Ramage, and P. Richtarik, "Federated Optimization: Distributed Machine Learning for On-Device Intelligence", arXiv:1610.02527 [cs], Oct. 2016.

[10] J. Choi, S.R. Pokhrel, "Federated Learning with Multichannel ALOHA", *IEEE Wireless Communications Letters*, early access, 2020.

[11] F. Ang, L. Chen, N. Zhao, Y. Chen, W. Wang, F. Richard Yu, "Robust Federated Learning with Noisy Communication", *IEEE Tr. on Communications*, early access, 2020.

[12] K. Yang, T. Jiang, Y. Shi, Z. Ding, "Federated Learning via Over-the-Air Computation", *IEEE Tr. on Wireless Communications*, early access, 2020.

10

Multi-Robot Systems

State-of-the-art robotics often deal with scenarios requiring multiple robots to complete a mission, say robots in factory automation as shown in Figure 10.1, a team of exploratory robots, or a platoon of autonomous driving vehicles, which introduces another important technology regarding *multi-robot systems* (MRS). Wireless communications and networking enables highly flexible and thus dynamic multi-robot systems with opportunities of further technology advance to form a *networked MRS*. A multi-robot system can be viewed as a *multi-agent system* (MAS) in AI, and we call a MAS of (wireless) networking as *networked MAS* (NetMAS).

Figure 10.1   Multiple robots together to assemble cars; photo from Forbes, https://www.fo rbes.com/sites/annashedletsky/2018/06/11/when-factories-have-a-choice-between-robots-a nd-people-its-best-to-start-with-people/#41e02d2e6d5f.

277

10.1 Multi-Robot Task Allocation

The immediate technology challenge for MRS is *multi-robot task allocation* (MRTA), which may be rooted from resolving the task complexity, enhancing overall performance of a MRS, increasing MRS's reliability, or collaboration among robots of different functionalities. The MRTA problem is therefore defined as: to find the task-to-robot assignments in order to achieve overall system goal(s)/mission(s), which has two sub-problems to accomplish:

(a) A set of tasks is assigned to a set of robots, or equivalently, a set of robots is assigned to a set of tasks.

(b) The behavior of multiple robots is coordinated in order to achieve the collective (usually cooperative or collaborative) tasks in an efficient and reliable manner.

Remark: The MRTA problem is essentially a dynamic decision problem, varying in time and phenomenon such as environment changes, flexible and dynamic orders, etc., which consequently implies MRTA is preferred having an iterative solution as time progressing.

10.1.1 Optimal Allocation

The MRTA problem can be straightforwardly formulated as an *optimal assignment* (OA) problem by defining $R = \{r_1, \ldots, r_m\}$ as a team of robots $r_i, i = 1, \ldots, m$, $T = \{t_1, \ldots, t_n\}$ as a set of tasks $t_j, j = 1, \ldots, n$, and $U = \{u_{ij}\}$ as the collection of robots' utility where u_{ij} denotes the utility of robot i executing task j. The objective of OA problem is to assign T to R, or *vice versa*.

Remark: The utility usually involves two aspects: the expected quality of task execution, Q_{RT} and the expected resource cost of execution, C_{RT}. Given the robot i is capable of executing task j,

$$u_{ij} = Q_{ij} - C_{ij} \qquad (10.1)$$

Generally speaking, the MRTA has to consider the following aspects:

- Single-task (ST) robots or multi-task (MT) robots
- Single-robot (SR) tasks or multi-robot (MR) tasks
- Instantaneous assignment (IA) or time-extended assignment (TA), while instantaneous assignment means the information regarding robots, tasks, and environments serves an instantaneous decision (or one-shot decision).

Definition (ST-SR-IA Optimal Assignment Problem): Given m robots and n tasks, while each robot is capable of executing the task (i.e. positive utility), the goal is to assign R (i.e. robots) to T (i.e. tasks) so as to maximize overall expected utility U.

The ST-SR-IA OA problem can be cast in many ways, typically as the well known *integral linear program*: to find mn non-negative integers α_{ij} that maximize

$$U = \sum_{i=1}^{m} \sum_{j=1}^{n} \alpha_{ij} u_{ij} \tag{10.2}$$

subject to

$$\sum_{i=1}^{m} \alpha_{ij} = 1, \ 1 \leq j \leq n \tag{10.3}$$

$$\sum_{j=1}^{n} \alpha_{ij} = 1, \ 1 \leq i \leq m \tag{10.4}$$

where α_{ij} functions like an indicator function to be either 1 or 0.

In other words, the ST-SR-IA OA problem can be understood as follows: given m robots, n tasks, with utility estimates for each of the mn possible robot-task pairs, assign at most one task to each robot. If such utilities can be centrally known to centrally execute linear programming, the optimal allocation time is $O(mn^2)$ in complexity.

Alternatively, a distributed auction-based approach can be used to find the optimal allocation, usually requiring time proportional to the maximum utility and inversely proportional to the minimum bidding increment. In order to understand such economically-inspired algorithms, the concept of linear programming duality is required. As all maximum linear programs, the OA problem has a dual minimum linear program, which can be stated as follows: to find m integers μ_i and n integers ν_j that minimize

$$\Psi = \sum_{i=1}^{m} \mu_i + \sum_{j=1}^{n} \nu_j \tag{10.5}$$

subject to

$$\mu_i + \nu_j \geq u_{ij}, \ \forall i, j \tag{10.6}$$

Remark: The Duality Theorem states that the original problem and its dual are equivalent, and that the total utility of their respective optimal solutions are the same.

Such optimal auction algorithm for task allocation usually works in the following way. A price-based task market is constructed, in which tasks are sold by imaginary brokers to robots. Each task j to be sold by a broker is placed a value c_j. Each robot i also places a value h_{ij} on task j. The problem then is to establish task prices p_j, which can determine the allocation of tasks to robots. To be feasible, the price p_j for task j must be greater than or equal to the broker's valuation c_j; otherwise, the broker would refuse to sell. Assuming that the robots are acting selfishly, the robot i will buy a task $t_{(i)}$ such that its profit is maximized.

$$t_{(i)} = \underset{j}{\operatorname{argmax}}\{h_{ij} - p_j\} \tag{10.7}$$

Such a market is said to be at equilibrium when prices are such that no two robots select the same task. At equilibrium, each individual's profit in this market is maximized. The two approaches (i.e., centralized and distributed) to solving the OA problem represent a tradeoff between computing time and communication overhead. Centralized approaches generally run faster than distributed approaches, but incur a higher communication overhead.

When the MRS consists of more tasks than robots, or if there is a model of task arrival process, then the robots' future utilities for the tasks can be predicted with some accuracy, and the problem is an example of ST-SR-TA OA problem. The following ST-SR-TA approximation algorithm works:

1. Optimally solve the initial $m \times n$ assignment problem.
2. Use the *Greedy algorithm* to assign the remaining tasks in an online fashion, as the robots become available.

Box: Combinatorial Optimization

The *combinatorial optimization* is developed in the theoretic framework based on *subset systems*.

Definition (Subset System): A subset system (E, F) is a finite set of objects E and a nonempty collection F of subsets, called independent sets, of E that satisfies the property that if $X \in F$ and $Y \subseteq X$ then $Y \in F$.

Definition (Subset Maximization): Given a subset system (E, F) and a utility function $U : E \to \mathbb{R}^+$, find an $X \in F$ that maximizes the total utility

$$U(X) = \sum_{e \in X} U(e) \tag{10.8}$$

> Given a maximization problem over a subset system, the canonical *greedy algorithm* is widely used to solve.
>
> **Proposition (Greedy Algorithm):**
> 1. Reorder the elements of $E = \{e_1, e_2, \ldots, e_n\}$ such that $U(e_1) \geq U(e_2) \geq \cdots \geq U(e_n)$
> 2. Set $X := \emptyset$
> 3. For $j = 1$ *to* n: if $X \cup \{e_j\} \in F$, then $X = X \cup \{e_j\}$

10.1.2 Multiple Traveling Salesmen Problem

That a robot consecutively executes tasks can be analog to the *traveling salesman problem* (TSP) in Chapter 2, as a robot corresponding to a salesman and tasks corresponding to visited cities. The nice aspect of this approach is to introduce distance measure corresponding to environment setting or time. MRTA is therefore corresponding to the *multiple traveling salesmen problem* (mTSP), by specifying m salesmen. These salesmen must cover all available nodes and return to their starting nodes (i.e. making a round trip for each salesman). The mTSP can be formally defined over a graph $\mathcal{G} = (\mathcal{V}, \mathcal{E})$, where $\mathcal{V}$ is the set of n nodes (i.e. tasks) and $\mathcal{E}$ is the set of edges (more precisely, directed edges to represent the order executing the tasks). Let $\mathbf{D} = [d_{kl}]$ be the distance matrix associated with $\mathcal{E}$. In general, asymmetric distance measure does not hold, that is, $d_{kl} \neq d_{kl}, \forall (k, l) \in \mathcal{E}$. By defining an indicator

$$\eta_{kl} = \begin{cases} 1, & \text{if edge } (k, l) \text{ is used in the trip} \\ 0, & \text{otherwise} \end{cases} \tag{10.9}$$

the mTSP can be formulated as follows:

$$\min \sum_{k=1}^{n} \sum_{l=1}^{n} \eta_{kl} d_{kl} \tag{10.10}$$

subject to

$$\sum_{k=2}^{n} \eta_{1,l} = m \tag{10.11}$$

$$\sum_{k=2}^{n} \eta_{l,1} = m \tag{10.12}$$

$$\sum_{k=1}^{n} \eta_{kl} = 1, l = 2, \ldots, n \tag{10.13}$$

$$\sum_{k=1}^{n} \eta_{kl} = 1, k = 2, \ldots, n \tag{10.14}$$

$$\eta_{kl} \in \{0, 1\}, \forall (k, l) \in \mathcal{E} \tag{10.15}$$

$$\sum_{k \in S} \sum_{l \in S} \eta_{kl} \leq | \, subTrip \, | - 1, \forall S \subseteq \mathcal{V} \setminus \{1\}, subTrip \neq \emptyset \tag{10.16}$$

Remark: There are many variants of mTSP to fit different scenarios of MRTA. However, NP-hard complexity is always the issue associated with this approach. Various computational algorithms have been developed for TSP and mTSP in literature.

10.1.3 Factory Automation

MRTA algorithms assign the tasks without considering the order of the tasks, while *crowdsourcing* is typically organized in this manner but not sure the reliability of task executions. However, the order of tasks does matter. Particularly, a special class of MRTA is about the *factory automation*, which assigns tasks following certain sequence(s). For example, in an automated production line to manufacture a product in massive scale, a task can be executed by the assigned robot only after another task by another assigned robot as shown in Figure 10.2, which brings in extra constraints into the optimization defined earlier.

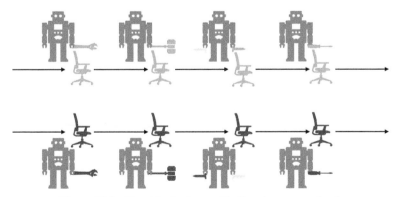

Figure 10.2   Two identical assembly lines in green and red.

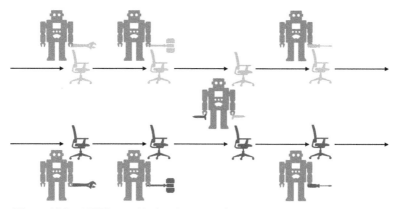

Figure 10.3 MRTA considering the assembly time for each robot by utilizing.

In addition to the order of tasks, another new dimension of MRTA in an automated assembly line is the execution time of a task. For example, the lth robot R_l executing task J_l takes time duration of τ_l, $l = 1, \ldots, L$. Up to now, τ_l, $l = 1, \ldots, L$ are treated as the same time duration, but generally and practically, they are not identical. For example in Figure 10.3, since the robot to screw nails can work two times faster than other robots, instead of 50% production capability being used, this robot can serve two production lines at the same time to save the factory installation cost and energy consumption.

An alternative to look into Figures 10.2 and 10.3 is via graphical approach. A robot can be defined as a state variable (i, j) that denotes the jth type-i robot. By treating robots as nodes and each movement in production order as a directed link/edge, the subsequent directed graph can fully describe the operation of entire multi-robot system. For exmaple, Figure 10.4 is illustrated to describe the MRTA in Figures 10.2 and 10.3. Similar to the traveling salesman problem (TSP), the assembly of a product is equivalent to visiting certain types of robots in order, while each robot corresponds to a city labelled (i, j).

Remark: Considering execution time of a sequence of robots can actually form a tandem queue according to queuing theory that is widely used in analyzing the performance of a network.

In recent years, beyond automated assembly line described above, *smart factory* in *smart manufacturing* has been suggested to revolution future industry, which is also known as *industry 4.0*. Instead of one product for an automated assembly line or even a factory in state-of-the-art manufacturing, smart manufacturing expects to rapidly respond the demand from the market

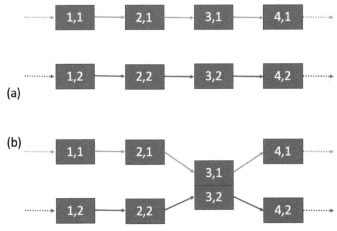

Figure 10.4 Equivalent graphs to (a) Figure 10.2 (b) Figure 10.3.

to flexibly arrange production flows for multiple products. Several stages are required to facilitate smart manufacturing:

(a) Plan what kind of products and corresponding quantity to manufacture, based on the market (online) analysis of supply-demand data that might be from Internet or an online mechanism.
(b) Acquire components and execute shipping logistics to smart factory through online methodology.
(c) Arrange the tasks to robots and determine energy-efficient production flows including moving the unfinished products among robots.

Stage (c) is of interest in this book and will be further in the following. Let us start from a simple example to demonstrate the superiority of smart factory aiming at flexible and efficient manufacturing.

Example (SR-ST MRTA for Smart Factory): A factory has 4 types of robots and 3 robots for each type as shown in (a) of Figure 10.5, which can form 3 automated assembly lines to manufacture desk as shown yellow in (b) of the Figure. In the mean time, these 4 types of robots can also be used to manufacture beds (in red) and sofas (in green) as (b) of Figure 10.5, which is not possible in today's fixed assembly lines until a good amount of time to re-setup and may strongly influence the production as original production. Even worse, in many cases, re-setup might not bring in financial benefits. For example, this factory can complete 3,000 desks each day (i.e.1,000 desks or beds/sofas per 4 robots per line), while manufacturing a desk can generate

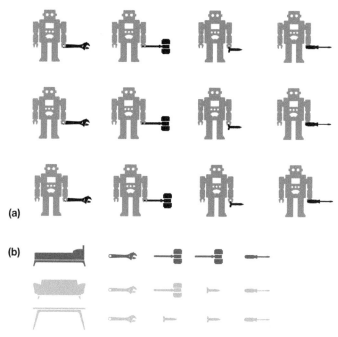

Figure 10.5 (a) Layout of robots in a factory (b) Possible products in flexible and smart manufacturing.

profit $100 and manufacturing a sofa or a bed can generate $120 profit. Mohsen is managing this factory and gets an order of 1,000 beds on the coming Thursday, which is not worth as he gives up production capacity for 2,000 desks per day since manufacturing a bed takes two hammer-robots, even ignoring the loss due to re-setup time.

With the advance of computing, control, and communications/networking, smart factory becomes possible to flexibly adjust the production by moving products in production among robots. Now Mohsen got another order of 1,000 sofas on the coming Thursday, and the smart factory can therefore manufacture 1,000 beds, 1,000 sofas, and 1,000 desks per day. The production can be easily found as Figure 10.4 with two possibilities (actually six but considering symmetry) as shown in Figure 10.6. The solutions can be found by modifying the graphical search in Chapter 2. Smart factory technology allows Mohsen to flexibly take these two orders with higher profit.

Another interesting problem is the energy-efficiency. In Figure 10.6, moving an unfinished product between different rows obviously takes more energy, say moving between rows 2 and 3 taking more energy than moving

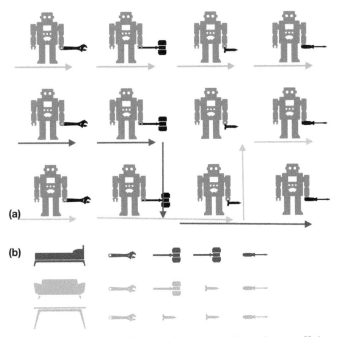

Figure 10.6   Production flows for three products to reach maximum efficiency of robot utilization.

between rows 1 and 3. Consequently, we mentioned earlier, what is the energy-efficient MRTA? It is easy to note that the sofa production executed in the row 1 (or row 3) as shown in 10.6 turns out to be the desirable robot assignment.

▶ **Exercise (Time-Sharing of A Robot):** In a smart factory, there are totally nine robots in four types of functionalities as shown in Figure 10.7. For each robot, the number of tools at both hands means capability of executing this number of tasks in a time slot, that is, a robot may execute multiple tasks in one time slot. The required tasks to manufacture a bicycle (in red) and a balance (in green) are also shown. The execution of tasks must follow the types, say a type-1 robot finishing the task for a product before delivering this unfinished product to type-2 (or type-3) robots in order. The cost to deliver to next neighboring robot is 1 for robots in the same row, and 2 in the same column. The diagonal delivery between two robots is therefore $\sqrt{1^2 + 2^2} = \sqrt{5}$, say from the robot in the first row and first column to the robot in the second row and second column. Please design the optimal production flow

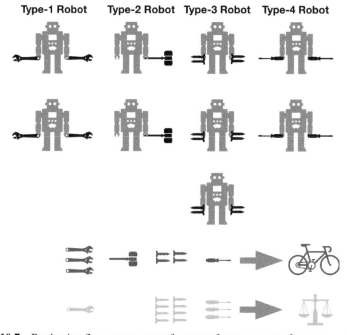

Figure 10.7 Production flow arrangement for smart factory to manufacture two products.

for this smart factory such that the production cost is minimum given that the throughput (i.e. the number of finished products) is maximum.

Remark: In above simple example of smart factory, wireless networking in next section is required to assign tasks to a multi-robot system and instruct mobile robots to move unfinished products along these flexible routes. In addition, queuing theory can be useful to model the MRS involving time-schedule, say time-sharing of a robot or different working durations for different robots.

10.2 Wireless Communications and Networks

When we developed MRTA in Section 10.1, it was treated as a pure computing problem. However, in real world, any multi-robot system involves information exchange relying on communication to complete the computing tasks. For example, in Section 8.3, two cleaning robots without communication consume pretty much the same time to complete the mission as a single robot since impossible to know each other's status. If the robots are mobile or have moving functionalities (such as movements of robot arms), wireless communication and networking technology has to apply, which we called *wireless robotics* as the title of this book.

10.2.1 Digital Communication Systems

A *communication system* is to bring a piece of information from one place (i.e. *transmitter*) to another (i.e. *receiver*) going through the medium to propagate (i.e. *channel*), while the information can be text, speech, audio, image, video, or the mixture of types of information. If the information is discrete or we quantize the information into discrete format, this is a *digital communication system*. Modern communication systems are always digital due to many reasons such as effective implementation in digital integrated circuits, easy to achieve spectral efficiency and better system performance (i.e. *bit error rate*, which is most common system performance measure and 10^{-6} is preferred for good quality of digital information transmission), inherent to facilitate error correcting codes and cryptography, etc.

As illustrated by Figure 10.8, a digital wireless communication system typically involves the following components or functional blocks.

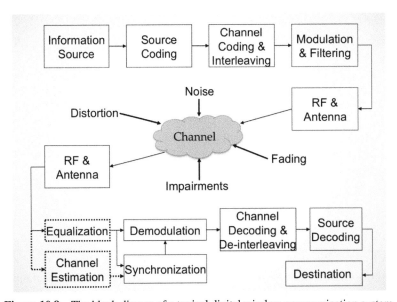

Figure 10.8   The block digram of a typical digital wireless communication system.

- Transmitter: As shown in the top portion of the figure before getting into the channel, a transmitter typically executes the following functions (i) to convert the analog information to digital information in minimal amount of bits with minimal possible distortion, that is, *source coding*; (ii) to amend redundant bits to protect digital symbols against errors, that is, *channel coding* (iii) to modulate information bits into digital symbols embedded into signal waveforms (i.e. *modulation*); (iv) through the RF section translating to carrier frequency and antenna(s), to transmit modulated and coded waveforms into the channel.
- Channel: The channel is primarily the wireless medium to propagate the waveforms, and also includes some effects from RF front-end and antennas. In addition to signal decay, the channel introduces (i) the embedded noise, which we usually consider *additive white Gaussian noise* (AWGN) whose properties have been briefly oriented in Chapter 4; (ii) signal distortion by the nonlinear effects from the wireless propagation, which may lead to *inter-symbol interference*; (iii) signal *fading* causing signal strength much lower than the expected level, typically resulted from multi-path propagation and large signal bandwidth; (iv) other impairments in the final baseband signal waveforms
- Receiver: The purpose of receiver is to reconstruct the signal from the received waveforms, and thus the receiver usually has much more complexity than the transmitter. The signal detection plays a core function at the receiver and sits in so-called *outer receiver*, which requires the assistance from so-called *inner receiver* to supply information derived from the received waveform and agreement between transmitter and receiver. Demodulation, channel decoding, and source decoding, are the primary functions at the out receiver. The outer receiver must accomplish a few key functionalities: (i) *synchronization* to align the received waveform with the transmitter in time, frequency, phase, and amplitude; (ii) *equalization*, to neutralize the inter-symbol interference; or (iii) *channel estimation*, to obtain the parameters of the channel such that the signal detection can be conducted in a smooth way.

Example (Digital Modulation): One of the core issues to design a digital communication system is to select an appropriate digital modulation scheme to embed information bit(s) into signal waveform. A waveform can be generally represented by the following mathematic equation.

$$A\sqrt{2}\cos(2\pi ft + \phi), 0 \geq t \geq T$$

where A denotes the amplitude; f denotes the (carrier) frequency; ϕ denotes the phase; T denotes the symbol period. Consequently, there are three fundamental digital modulation schemes:

- Amplitude Shifted Keying (ASK) by embedding information into the amplitude. For example, $A_m\sqrt{2}\cos(2\pi ft + \phi), m = 0, 1$ represents possible binary signals $\{0, 1\}$.
- Frequency Shifted Keying (FSK) by embedding information into the frequency. For example, $A\sqrt{2}\cos(2\pi f_m t + \phi), m = 0, 1$ represents possible binary signals $\{0, 1\}$, where $|f_1 - f_0| \gg 1/T$.
- Phase Shifted Keying (PSK) by embedding information into the amplitude. For example, $A\sqrt{2}\cos(2\pi ft + \phi_m), m = 0, 1$ represents possible binary signals $\{0, 1\}$. When $\phi_0 = 0$, $\phi_1 = \pi$, this is known as binary PSK (BPSK), which generates the largest possible signal separation to against AWGN and is also known as *antipodal signal*. BPSK is widely used in modern digital communication systems. If we implement BPSK onto in-phase channel (i.e. I-channel) and another BPSK in parallel onto quadrature-phase channel (i.e. Q-channel), this is called quadrature PSK (QPSK), which is most efficient in bandwidth and error rate performance by transmitting two information bits at a time.

Further hybrid above mechanisms could be possible. For example, to develop a modulation for high bandwidth application (i.e. high rate transmission), we can use 16 quadrature-amplitude modulation (16-QAM), 64-QAM, or even 256-QAM. Figure 10.9 illustrates the signal constellation of QPSK and 16-QAM in signal space, where QPSK can carry two information bits (i.e. 00, 01, 11, 10) and 16-QAM can carry four information bits.

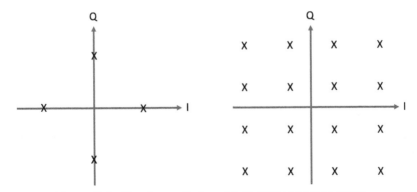

Figure 10.9   Signal constellation of (left) QPSK (right) 16-QAM.

Example (Synchronization): By ignoring the noise, the received waveform can be generally represented as

$$y(t) = \sum_n \alpha(t) e^{i\theta(t)} s(t - nT - \tau) \qquad (10.17)$$

- To deal with large dynamic range of $\alpha(t)$, we usually adopt *automatic gain control* (AGC), which can be facilitated by operational amplifier in electronics.
- To recover τ, this is known as *timing recovery*. Signal waveforms are usually based on symbols or bits, and such timing recovery is also called symbol synchronization or bit synchronization.
- To recover $\theta(t)$ is known as *carrier recovery*, which consists of *frequency estimation* and *phase estimation*.
- The alignment of n is known as *frame synchronization* or block synchronization, which usually plays an important role in multimedia communication.

Synchronization mechanisms can be usually realized either by estimation or by M-ary hypothesis testing. For example, We may consider such symbol (or actually bit) synchronization as selecting the most probable timing delay among $\{\tau_1, \ldots, \tau_N\}$, N hypotheses of time delay in baseband timing. A typical case is to consider a uniform sampling within a symbol/bit period, that is, to assume $\tau_n = (n-1)T/N, n = 1, \ldots, N$. This is actually an N-ary hypothesis testing (please refer Chapter 4) to select a hypothesis time delay closest to the actual timing.

$$H_n : \tau_n \ closest \ to \ actual \ timing \ n = 1, \ldots, N$$

In other words, from N-ary hypothesis testing in AWGN, the timing is determined by

$$\hat{\tau} = \underset{n}{\operatorname{argmax}} \int_0^T y(t, \tau) s(t - \tau_n) dt$$

by assuming that carrier recovery and amplitude control are done.

Remark (Channel Capacity): C. Shannon innovated well known *information theory* to explore the fundamental limits of communication systems. For each channel, there exists a *channel capacity*, C. If the transmission rate over this channel is R and $R \leq C$, there exists a way to achieve reliable communication, which is known as *channel coding theorem*. For an AWGN channel of bandwidth W and signal-to-noise-ratio (SNR), then

$$C_{AWGN} = W \log_2(1 + SNR) \qquad (10.18)$$

Remark: In general sense, even copying of genetic information in cell splitting can be viewed as a sort of digital communication. Information exchange among multiple agents can always be treated as a communication system or networks, to form a networked multi-agent system (MAS).

10.2.2 Computer Networks

A digital communication system typically involves one transmitter and one receiver, that is, point-to-point communication. Hereafter, we have to consider the communication among multiple users or nodes, to form a communication network. In particular, if each node has the computing, storage, and forward/receive capability, these nodes can form a *computer network*. However, please note that a special nature of communication and network industry is to develop communication devices for agents, and these devices must be inter-operable, even from different vendors. To accomplish such nature in diverse applications, International Organization for Standardization (ISO) develops a lot of standards for computer networks and thus Internet.

Figure 10.10 depicts the 7-layer *open system interconnection* (OSI) structure of computer networks. Such a 7-layer partition might not be good to optimize network efficiency. However, it is of great value to implement large

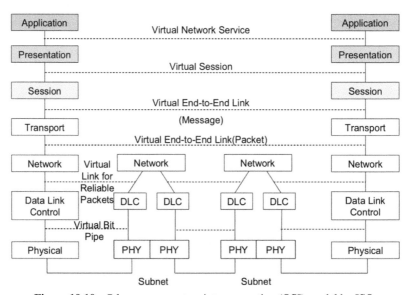

Figure 10.10 7-layer open system interconnection (OSI) model by ISO.

scale of networks via such a layered-structure. Engineers can implement a portion of software and hardware in a network independently, even plug-in networks or replace a portion of network hardware and/or software, provided that the interfaces among layers and standards are well defined. Considering the nature of *stochastic multiplexing* packet switching networks, OSI layer structure may easily promote quick progress of computer (wireless) networks and wireless networking for robotics.

All upper 4 layers are mainly "logical" rather than "physical" concept in network operation; while physical signaling is transmitted, received, and coordinated in lower two layers, physical layer and data link layer. Physical layer of a wireless network is thus to transmit bits and to receive bits correctly in wireless medium, while medium access control (MAC) is to coordinate the packet transmission utilizing the medium formed by a number of bits. The data link layer has two major functions: logic link control (LLC) and medium access control (MAC). The network layer is to utilization of network resources, routing, flow control, etc. For mobile robots, mobility management and subsequent radio resource management of wireless networking shall be handled by the network layer. Transport layer and session layer correspond to virtual end-to-end delivery of packets and messages respectively.

Modern wireless networks or mobile networks can be categorized into two classes: *infrastructured* and *ad hoc* networks. Each infrastructured wireless network as shown in Figure 10.11 has a (high-speed) backbone (wired or wireless) network to connect a number of base stations (or access points). The mobile stations communicate through the base stations then the backbone network and on toward the destination mobile station. The packet delivery relies on an infrastructure consisting of the backbone network and base stations. On the other hand, a number of mobile stations may establish an ad-hoc network without any infrastructure, as in Figure 10.11, where each link between two nodes (i.e., mobile stations) is plotted and these links build up the network topology of an ad-hoc network. Although ad hoc networks have been considered in robotics, its technical challenge about *scalability* has been mostly overlooked, which give a technical opportunity in wireless robotics.

An immediate problem associated with wireless robotics is the routing in a *mobile ad hoc network* (MANET), which is described in Section 2.3. Bellman-Ford algorithm and Dijkstra algorithm well serve the purpose of routing in an ad hoc network. Mobility makes these algorithms more challenging to execute as frequent update routing table for each node is required. Another way to assist is clustering the MANET with speaker selection algorithm for each cluster.

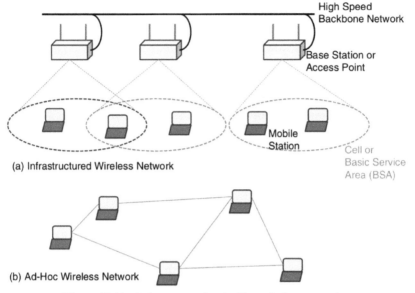

Figure 10.11   Infrastructured and ad hoc wireless networks.

10.2.3 Multiple Access Communication

Recalling the layered network structure, we need a sub-layer called *medium access control* (MAC) between the DLC and physical layer. The purpose of this extra sub-layer is to allocate the multi-access medium various nodes. The method to coordinate physical transmission among various nodes in a computer/ communication network is known as a multiple access protocol, which also serves as the essential function in wireless networks.

The pioneering multiple access system would the ALOHA, initially for a satellite communication (actually information collection) system with a topology of multi-point to single-point. Nodes (earth stations) on the ground are trying to access the satellite to relay the packets and thus require a multiple access protocol to coordinate the transmission in a distributed way. When a set of nodes simultaneously share a communication channel, the reception is garbled if two or more nodes transmit simultaneously, which is known as collision. And, the channel is unused (or idle) if none transmit. The challenge of multiple accesses (also multi-access) is how to coordinate the use of such a channel through a distributed or centralized way. In this section, we focus on distributed multiple access protocol family, which is widely used in wireless data networks. Pure ALOHA is simply: (i) when a node has a packet to transmit, it transmits; and (ii) the node listens to the channel. If collision happens, the node re-schedules transmission of the packet by a (random) backlog algorithm. Otherwise, the node transmits the packet successfully.

To study the multiple access protocol, we usually consider the time axis to be slotted for node operation. Assumptions for the idealized slotted multi-access model are summarized as follows:

- Slotted system: All transmitted packets have the same length and each packet requires one time unit (called a slot) for transmission. All transmitters' are synchronized so that the reception of each packet starts at an integer time and ends at the next integer time.
- Poisson arrival: Packets arrive for transmission at each of the m transmitting nodes according to independent Poisson processes with l/ m arrival rate.
- Collision or perfect reception: The packet is received either in a perfect way or in collision to lose information.
- $\{O, 1, e\}$ Immediate feedback: The multiple access channel can provide feedback to distributed nodes with three possibilities $\{0, 1, e\}$, where 1 stands for successful packet transmission and reception; 0 stands for channel idle with no packet transmission; and e stands for collision(s) in the multiaccess channel.
- Retransmission after collisions: Each packet involved in a collision must be retransmitted in some later slot, with further possible such retransmission until a successful transmission. A node with a retransmission packet is called backlogged.
- (a) No buffer in each node. (b) Infinite number of nodes in the systems.

▶ **Exercise:** Please show that the pure ALOHA has throughput (i.e., average number of packets successfully transmitted per packet transmission time) of $1/(2e)$, where e is Euler constant.

An obvious drawback of pure ALOHA is any collisions to possibly last for two packet periods. An immediate improvement is for all active nodes to transmit the packets at the beginning of each time slot, which confines any collision within one packet period, which is called *slotted ALOHA* and widely used in wireless networks.

▶ **Exercise:** Please show that the slotted ALOHA has throughput (i.e., average number of packets successfully transmitted per packet transmission time) of $1/e$ as Figure 10.12, which is just near 37%.

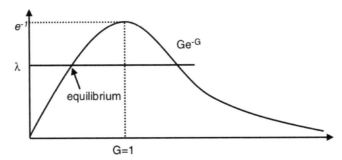

Figure 10.12   Throughput of slotted ALOHA.

As a matter of fact, if nodes can collect some kind of information from the multiple access channel, it intuitively results in better performance multi-access protocols. The most straightforward approach might be *carrier sensing*, in which a node listens to a possible transmission in the channel first, and then transmits. Consequently, carrier sensing can be called *listen-before-transmission* (LBT). The multiple access protocols executing carrier sensing functionality prior to transmission are known as *carrier sense multiple access* (CSMA), which is adopted in the IEEE 802.11 wireless local area networks (WLANs) that is also called WiFi.

10.3 Networked Multi-Robot Systems

After introducing the wireless communications and networking, it is expected beneficial to apply wireless communication technology to multi-robot systems. However, this subject has only lightly been studied in literature. In the following two sub-sections, two illustrations of multi-robot systems in earlier chapter will be explored to see the benefits of apply wireless technology.

10.3.1 Connected Autonomous Vehicles in Manhattan Streets

In Section 5.2.3, RL is used to model the navigation of autonomous vehicles (AVs) in the Manhattan streets. One major cause to lose the efficiency of road utilization comes from the possibility of two or more AVs getting into an intersection at the same time, which requires the rule of 4-way stop sign (i.e. first-come-first-go). Please note that it is possible to consider the AVs in the region of interest as a *multi-agent system* (MAS). It is straightforward to employ wireless communication technique to connect AVs and alleviate the congestion in the intersections, if two (or more) AVs can foresee their potential congestion in front and re-route their navigation paths to avoid waiting in the intersection. In such a MAS, since each AV knows the street map, agents (i.e. AVs) can exchange their reward maps and policies to dynamic adjust RL.

Ideal wireless communication

To serve as a bench mark, we start to explore the impacts from ideal communication (error-free, unlimited radio/channel resources with perfect coordination, contention free, with the communication range for all involved agents) on the MAS of RL. No matter in communication mode or not, each agent needs to recognize or predict the other agents' future movements (policies) in order to avoid crashing. At time k the agent recognizes other vehicle and generates reward map $R_{i,k}$, but from $k + d, d = 1, 2, ...$ will be the expected reward map $\hat{R}_{i,k+d} = [\hat{r}_{s_{k+d}}]$, d is length of next time step. If the horizon depth of policy is D, the i-th vehicle generates predicted reward map $\mathbb{R}_{i,k:k+D} = \{R_{k,i}, \hat{R}_{i,k+1}, \ldots, \hat{R}_{i,k+D}\}$. Due to wireless communication for information exchange, we modify RL by incorporating information from another agent as follows.

- With the aid of wireless communication, the i-th vehicle recognizes the position of j-th vehicle and gets the j-th vehicle's reward map $R_{j,k}$
- The i-th vehicle gets the policy of the j-th vehicle's or predicts the future movement, the i-th vehicle updates own reward map $\mathbb{R}_{i,k:k+D}$ with $\mathbb{R}_{j,k:k+D}$.
- Similarly, the i-th vehicle also share own reward map $\mathbb{R}_{i,k:k+D}$ to the j-th vehicle
- Q-learning proceeds based on the modified reward $\mathbb{R}_{i,k:k+D}$

The agent calculates and sets the reward at each state at s_k, and generates the update reward map of depth D. We already defined r_{s_k} at (1) in Section 5.2.3. In the communication mode, the reward map is used to compute the policy, while it represents the reward based on the predicted movement without communication mode.

No matter in the communication mode or not, basic procedure of reward updating is as follows. Denote $\mathbb{I}_k$ as the collection of vehicles that i-th vehicle observed or communicated at time k.

$$\mathbb{R}_{i,k:k+D} \leftarrow \mathbb{R}_{i,k:k+D} \bigcup_{j \in \mathbb{I}_k} \mathbb{R}_{j,k:k+D} \qquad (10.19)$$

With updated expected reward map $\mathbb{R}_{i,k:k+D}$, the agent calculates Q-values for the i-th vehicle $Q_{i,k}(s,a)$, s is a certain state $s = \{l_x, l_y\}$, $l_x = 1, 2, \ldots, L_x$, $l_y = 1, 2, \ldots, L_y$ and a is possible action from $\{forward, left, right, stay\}$ at state s. The consequent Q-learning is

$$Q_{i,k+1}(s,a) \leftarrow Q_{i,k}(s,a) + \alpha[r_{s'} + \gamma \max_a Q_{k+1}(s',a) - Q_k(s,a)] \quad (10.20)$$

The updated policy of the i-th vehicle, $\pi' = \{a_{k+1}, \ldots a_{k+D}\}$, can be derived by the updated Q-values, and iterating updated $\mathbb{R}_{i,k+1:k+D+1}$ and then swapping the policies with other vehicles. The basic updating is the same procedures for each mode (with communication or without communication), but the procedures of generating rewards map $\mathbb{R}_{i,k:k+D}$ depends on the utilization of communication or not. Algorithm 1 summarizes such new procedure of RL. $\mathbb{R}_{i,k:k+D,com}$ presents the reward map for the communication mode, and $\mathbb{R}_{i,k:k+D,obs}$ is for the reward map for observation map (without communication). Without communication mode, the agent utilizes the portion of previous reward map $\mathbb{R}_{i,k-1:k+D-1,com}$ from the communication. Consequently, it is possible for each agent to generate rewards map $\mathbb{R}_{\mathbb{I}_k}$ and to share with other agents.

V2V communication

In wireless communications, direct communication between two vehicles is known as vehicle-to-vehicle (V2V) communication. Similarly, if there is communication infrastructure, say optical fiber backbone network, the uplink communication from a vehicle to the infrastructure is known as vehicle-to-infrastructure (V2I) communication, and the downlink communication from the infrastructure to vehicles is known as infrastructure-to-vehicle (I2V) communications. The scenario that ideal wireless communication is applied to V2V communication as shown in the left side of Figure 10.13 represents the simplest case to study.

- Within the communication range r, the i-th vehicle recognizes other vehicles' (j's-th vehicle, $j \in \mathbb{I}_k$) states.
- Position as well as the expected reward map $\mathbb{R}_{j,k:k+D}$ of another vehicle can be obtained.
- Each vehicle has infinite channel resource such that V2V communication can be retained as real-time (for RL).

The V2V communication proceeds as follows:

Algorithm 1: Q-learning with modified rewards

---

1 <u>function</u> $Q_{k+d}(s,a), R_{i,k}, \mathbb{R}_{i,k:k+D,com}, \mathbb{R}_{i,k:k+D,obs}, s \in \mathcal{S}, a \in \mathcal{A}$;

2 Initialization $k = 0, d = 0, R_{i,k} = 0, Q_{k+d}(s,a) = 0, \mathbb{R}_{i,k:k+D,com} = 0, \mathbb{R}_{i,k:k+D,obs} = 0$

3 **for** *k until Vehicle i at the destination* **do**

4   Observe r_{s_k} for $R_{i,k}$, s'(Possible next state) from s_k with taking action a_k;

5   Update $R_{i,k}$;

6   **if** *There are communication (V2V or V2I2V)* **then**

7   **for** $d \leftarrow 0$ **to** D **do**

8              $\mathbb{R}_{i,k:k+d,com} \leftarrow \mathbb{R}_{i,k:k+d,com} \bigcup_{j\in\mathbb{I}_{k+d}} \mathbb{R}_{j,k:k+d,com}$;

9   **for** *each state s* **do**

10   Derive r_s from $\mathbb{R}_{i,k:k+d,com}$ $Q_{k+d}(s,a) \leftarrow Q_{k+d}(s,a) + \alpha[r_{s'} + \gamma \max_a Q_{k+l+1}(s',a) - Q_{k+l}(s,a)]$

11   **end**

12   **end**

13   **else**

14   **for** $d \leftarrow 0$ **to** D **do**

15              $\mathbb{R}_{i,k:k+d,obs} \leftarrow \mathbb{R}_{i,k:k+d,obs} \bigcup \mathbb{R}_{i,k-1:k+d-1,com}$;

16              $\mathbb{R}_{i,k:k+d,obs} \leftarrow \mathbb{R}_{i,k:k+d,obs} \bigcup_{j\in\mathbb{I}_{k+d,obs}} \mathbb{R}_{j,k:k+d,obs}$;

17   **for** *each state s* **do**

18   Derive r_s from $\mathbb{R}_{i,k:k+d,obs}$ $Q_{k+d}(s,a) \leftarrow Q_{k+d}(s,a) + \alpha[r_{s'} + \gamma \max_a Q_{k+l+1}(s',a) - Q_{k+l}(s,a)]$

19   **end**

20   **end**

21   **end**

22   ;

23   **end**

24 **end**

- i-th vehicle probes other vehicles' reward maps at time k if any ($j \in \mathbb{I}_k$) vehicle within the communication range, the i-th vehicle thus receives $\mathbb{R}_{j,k:k+D}$ successfully
- No matter receiving from other vehicles or not, the i-th vehicle broadcast own reward map $\mathbb{R}_{i,k:k+D}$ and $\mathbb{R}_{j,k:k+D}$ according to pre-determined rule of order.

The basic principle of the updating reward map is the same as (8). Via V2V communication, the agent receives information from other vehicles, but also shares information with others. Suppose $\mathbb{I}_k$ to be the collection of vehicles IDs, which the i-th vehicle communicates at time k. When the agent

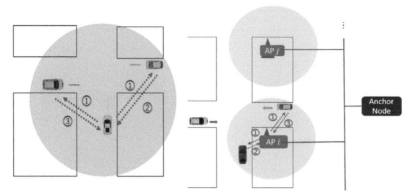

Figure 10.13   V2V communication (left): Each vehicle can communicate with other vehicles within the communication range, V2I2V communication (right): Each vehicle can communicate to APs within radio communication range. Anchor node (AN) is expected to equip edge computing and govern the operation of APs.

received other vehicles' reward maps for learning,

$$\mathbb{R}_{i,k:k+D} \leftarrow \mathbb{R}_{i,k:k+D} \bigcup_{j \in \mathbb{I}_k} \mathbb{R}_{j,k:k+D} \qquad (10.21)$$

Other vehicles also have multiple vehicles' policies, the i-th vehicle can update $\mathbb{R}_{\mathbb{I}_k}$. For other vehicles obtain the i-th vehicle's policy, then omit $\mathbb{R}_{i,k}$.

$$\mathbb{R}_{j,k:k+D} \leftarrow \mathbb{R}_{j,k:k+D} \bigcup \mathbb{R}_{i,k:k+D} \qquad (10.22)$$

V2I2V communication

It is well known that ad hoc networking (e.g. V2V communication) suffers from scalability problem. We therefore examine V2I2V, two-hop wireless communication as the comparison, to figure whether roadside infrastructure is useful. The network infrastructure can support high-bandwidth low-latency communication. To be fair in comparisons, the communication range of V2I and I2V is set as half of that in V2V communication. Suppose access points (APs) are placed in every corner ob the block, and each vehicle has the communication range r. If the vehicle finds AP in the communication range, then the vehicle connects to AP and swap the reward map with others.

- Within the communication range r, the vehicle connects to an AP
- The vehicle sends the information about the position and the policy of future movement

- Each vehicle and AP have infinite channel resource unit (RU) such that they can communicate in real-time (before next time instance)
- Once an AP collects vehicles' information, then sends network infrastructure to broadcast to other vehicles via other APs (in two time instances due to V2I and I2V)

V2I2V, though in two-hops, enjoys one advantage of sharing reward map to more vehicles even outside direct communication range, thanks to network infrastructure. The network infrastructure (NI) relays the reward maps $\mathbb{R}_{AP_m}$ through APs $m \in M$, and send these reward maps to the all vehicles by other APs. The policies are updated in the following manner.

$$\mathbb{R}_{AP_{m,k:k+D}} \leftarrow \bigcup_{i \in M_\neg} \mathbb{R}_{i,k:k+D} \tag{10.23}$$

$$\mathbb{R}_{NI_{k:k+D}} \leftarrow \bigcup_{m \in M} \mathbb{R}_{AP_{m,k:k+D}} \tag{10.24}$$

$$\mathbb{R}_{i,k:k+D} \leftarrow \mathbb{R}_{AP_{m,k:k+D}} \leftarrow \mathbb{R}_{NI_{k:k+D}} \tag{10.25}$$

Multiple Access Communications

Above study proceeds on an important assumption, ideal communication that all communication packets are perfectly coordinated by a genie. Practically, there exist finite Radio Resource Units (RRUs), and then certain random access protocols in noisy communication channel(s) must be employed to exchange information among AVs, which implies installation of random access over multi-agent system, as a very unique investigation on multi-agent system (MAS). Suppose there is one multiple access communication channel is available and AVs/agent are using slotted ALOHA as benchmark.

The slotted ALOHA system operates is defined as in the previous section.

- Time is segmented into slots of a fixed length for transmission
- When a node has a packet to transmit, it waits until the start of next slot to transmit
- The node listens to the channel, and if there are no collisions during the slot, it transmits packet successfully
- If collision happens, the node schedules another re-transmission by backlogging

Obviously, slotted-ALOHA, as a random access suffering potentially large dealy can not deal with a highly dynamic system that requires to collect information for real-time decision making in ultra-low latency, such as in just several milliseconds for an AV [5]. Re-transmission of a message may be useless even being correctly received later. Therefore, modification of slotted ALOHA is required to support RL in this scenario, and named as *real-time ALOHA* (rt-ALOHA), which aligns the design trend of ultra reliable and low-latency communication (uRLLC) by grant-free access, no acknowledgement, and discarding re-transmission [5].

- There is no retransmission and thus backlogging
- When the channel is busy, the agent (i.e. AV) is ready to receive immediately
- When the channel is idle, the agent broadcasts the message, and ready for receiving from others immediately after transmission without any acknowledgement by the receiving agent(s).

V2V communication by rt-ALOHA

Applying the slotted rt-ALOHA, each agent/AV adopting realistic V2V communication executes the following operation procedure with reference of Figure 10.13:

- When the channel is idle,
    1. If the channel is idle, the vehicle broadcasts $\mathbb{R}_{i,k:k+D}$ after random back-off time (① in the left part of the Figure 10.13)
    2. After broadcasting, the vehicle will wait for receiving the other vehicle's reward map $\mathbb{R}_{j,k:k+D}$
    3. Even though the receiving is success or not, the vehicle goes back to the channel sensing for the next broadcasting

- When the channel is busy,
    1. If the channel is busy, the vehicle will wait for receiving the other vehicle's reward map $\mathbb{R}_{j,k:k+D}$ (②, ③ in the left part of Figure 10.13)
    2. Even though the receiving is success or not, the vehicle goes to the channel sensing

Unlike the original ALOHA protocol, there are no backlogging process nor re-transmission process because every vehicle is keep moving and the environment is rapidly changing, therefore if the communication does not success, the information swapping will fail. The failure of the communication means the vehicle applying stop-and-see mode in Section 5.2.3 (no communication mode). Channel Sensing and broadcasting or receiving the packets is executed in one time step.

V2I2V communication by rt-ALOHA

In the multiple access communication, more collisions occur with the large communication range of V2V communication. For the ideal case, the wider range helps vehicles to get more information, however, in the multiple access, the communication failures due to collisions or multi-access interference (MAI) occur more frequently, which leads poor or the similar performance with non-communication mode for the delay. From this aspect, V2I2V helps vehicles to have other vehicles' information with smaller chance of packet collisions, which is adopted to examine the benchmark performance. Since V2I2V involves two segments of wireless transmissions, smaller radio range to eliminate the probability of collisions in radon access is expected.

Similarly to the V2V communication with rt-ALOHA, the operation of rt-ALOHA V2I2V communication proceeds as follows:

- If the channel is idle, AP broadcasts $\mathbb{R}_{AP_n\downarrow}$ (① in the right part of Figure 10.13)
- If AP is within the vehicles' communication range and vehicles receive $\mathbb{R}_{AP_n\downarrow}$ from AP, they try to prepare to broadcasting
- After random back-off time each vehicle starts to broadcast $\mathbb{R}_{i,k:k+D}$ (②, ③ in the right part of Figure 10.13)
- Even though the receiving is success or not, the AP goes to the channel sensing for broadcasting

Unlike the V2V communication, each vehicles keeps driving with waiting for APs broadcasting. Once receiving the $\mathbb{R}_{AP_m,k:k+D}$, each vehicle recognizes that AP is within the communication range and the vehicle can connect to the AP, and AP receives $\mathbb{R}_{AP_m,k:k+D}$ from cars successfully, then sends to AN and received the updated $\mathbb{R}_{AN_{k:k+D}}$. The possible situation of collisions is when the multiple vehicles try to transmit at the same time because they are out of communication range of each other but within the communication range of AP.

Simulations

Based on $X \times Y$ block Manhattan Street Model in Figure 10.13, $X = 4, Y = 6$, and the length of block $b = 5$, we simulate AVs driving to the destinations with Reinforcement Learning. The following simulations are about average delays by vehicle's arrival rate with different communication pattern. The average delay (extra step) is the delay from the minimum steps to the destination (i.e. the difference from the number of steps when only 1 car in the street). From formula (10) for calculate expected rewards for observation mode, we set the probability of vehicle staying at the same state $p_{stay} = 1/2$ because, when the observed vehicle in front of the intersection, the vehicle cannot recognize the vehicle will be move forward or stay at the same state, and also set p_a uniform distributed, which is the probability of the vehicle take action a. At each state p_a is uniform distributed by the number of possible action of vehicle at each state. We set arrival rate λ which is the average number of vehicles coming into the map at 1 time step. And the simulation result is from $k = 70$ to $k = 300$ because we need to wait the number of vehicles is stable and $k = 70$ is the time which the number of vehicles does not vary for $\lambda = 0, \ldots, 10$.

Average delay for different communication Pattern: We simulated the average extra delays (steps) by the vehicles' arrival rate into the map with different mode (without communication, ideal V2V communication, ideal V2I2V communication). The horizon depth of learning is $D = 5$, and there are no packet collisions and no interference with infinite channel.

Figure 10.14 shows the how the communication enhances RL with decreasing average extra time step. Without communication applies the observation mode, and V2V and V2I2V apply ideal communication (infinite channel and no interference). V2V with larger range and V2I2V with smaller range show the similar performance because V2I2V helps to get more

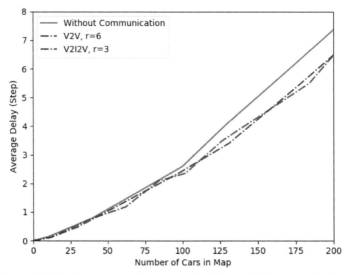

Figure 10.14   Average extra time step with different ideal communication.

rewards maps through the AN. The corresponding number of vehicles during stable state for arrival rate is 40 for $\lambda = 1$, 85-90 for $\lambda = 2$, 100-105 for $\lambda = 2.4$

Communication with packet loss: For both V2V and V2I2V communication with infinite channel bandwidth, we focus on the random packet loss or packet drop with rate $p = 0.01, 0.1, 0.3$. By this way, the communication will not be perfect even though vehicles are connected within their communication range. There are packet errors or packet drops with probability p. Figure 10.15 shows the results of average delay with different packet loss rate p for V2C communication and V2I2V communication. For V2V, the communication range is set $r = 6$, and for V2I2V, the communication range is set $r = 3$; With shorter range in V2I2V, each vehicle can have as more information as the larger range with V2V. The influence of the packet-loss rate affects when the large number of cars are in the street.

rt-ALOHA: Assuming the slotted rt-ALOHA for V2V and V2I2V communication, there exists only 1 access channel. When the vehicle sense the channel is idle it broadcasts the reward map after the random back-off time $(0 - 15,$ and which is relatively small than the duration of slot time). When the multiple vehicles try to broadcast at the same time in case the same random back-off time, a collision happens to result in transmission failures.

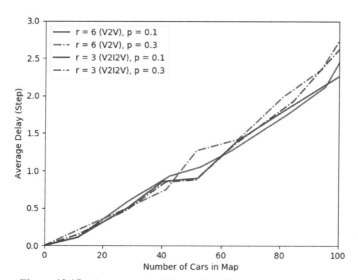

Figure 10.15 Average extra delays with different packet error rate.

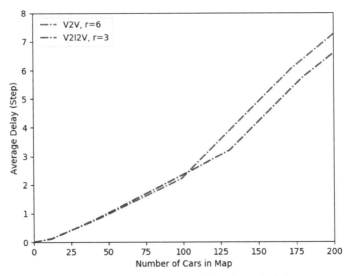

Figure 10.16   Communication with 1-channel ALOHA.

Figure 10.14 shows the V2V with communication range $r = 6$ and V2I2V with communication range $r = 3$ in the ideal situation. Figure 10.16 shows the same parameters for V2V and V2I2V with ALOHA protocol.

As the number of AVs increasing, more communication failures happen and less congestion can be consequently resolved, either using V2V or V2I2V, because each vehicle can only use single-channel and the collisions may occur. If more than 2 vehicles simultaneously broadcast own reward maps, the collisions occur. V2I2V communication of smaller range is better performance than V2V, which results in the smaller collision probability than V2V, but shows the same performance with V2V in ideal communication. Generally, speaking, wireless communication indeed improves the performance of mobile robots (and AVs) by intuitively utilizing machine learning information, at least in this resource-sharing MAS (i.e. multiple agents use the same resource by knowing public reference/map in advance). We will look another illustration of specular application interest in mobile robotics.

10.3.2 Networked Collaborative Multi-Robot Systems

As illustration of a single mobile service robot to clean a designated floor (Figure 8.9) was introduced in Section 8.3, which involved RL and planning for navigation. An interesting application scenario is to deploy multiple

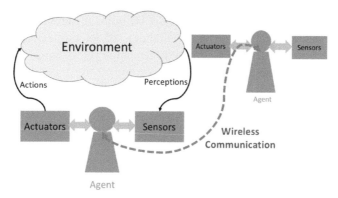

Figure 10.17 A collaborative two-robot system with wireless communication.

collaborative robots to form a collaborative multi-robot system (MRS), while Figure 10.17 illustrates this more complicated MRS or MAS.

Please note that this collaborative MRS is different from the resource-sharing MRS in earlier sub-section. Each agent/robot in the resource-sharing MRS/MAS has its own mission to accomplish given a common public reference (say, street map), but collaborative MRS has a common goal among robots without any (precise) public reference available. Any robot in a collaborative MRS must execute its (common) goal by creating private reference (e.g. robot pose). Therefore, what is the effective way for a collaborative MRS (or MAS) to operate.

- As Figure 10.20, two robots without (wireless) communication and thus no information exchange, can not save much time over a single robot to complete the mission. Robots or intelligent agents suffer a sort of *tragedy of commons* without proper information exchange. Wireless communication or robotic communication is therefore so vital to design multi-agent systems.
- A robot of reinforcement learning relies on appropriate planning and localization to achieve efficiency. Once multiple robots collaboratively work toward a common mission without public reference (i.e. global floor map in this illustration), exchange of private reference from individual experience can be very useful, in addition to exchange of reward-action map indicated from the previous sub-section or [6].

In this sub-section, wireless communication will be developed in three aspects: (i) what to communicate among collaborative agents since such machine-to-machine communications would be quite different from well-known personal communications (ii) advantages to adopt wireless communication in collaborative NetMAS (iii) how to design wireless communication functionalities for collaborative agents.

Ideal communication

First assume the agents communicate under ideal conditions, that is, infinite bandwidth, error-free transmission, and perfect coordinated multiple access among agents. The only factor in concern is the communication range. Suppose there are N robots operating independently, and each robot equipped with wireless communication. At the time a robot transits to the next state, it searches for other agents within range r. For example, at time t two robots u_i and u_j are at location $v_{ab} = (a, b)$ and $v_{pq} = (p, q)$ respectively. As the state of a robot is defined as $y_t^i = v_{ab}$, then any u_j satisfies $\overline{y_t^i y_t^j} = \sqrt{(a-p)^2 + (b-q)^2} \leq r$ can communicate with u_i.

If more than two robots are able to communicate, say u_i, u_j, u_k, they can simultaneously communicate with the two others respectively, i.e. mutual transmissions between three pairs: u_i and u_j, u_j and u_k, u_k and u_i are allowed to happen at the same time. Briefly saying, in the scenario of ideal communication, any two agent can transmit and receive packets perfectly as long as they lie inside others' communication range. $\forall i, j \in \mathbb{N} : 1 \leq i, j \leq N, i \neq j, Pr\left\{u_j \text{ receives from } u_i \middle| \overline{y_t^i, y_t^j} \leq r\right\} = 1$.

Information exchange and integration

To comprehend what to communicate among collaborative agents, an agent's private reference (map) can be very helpful to other agents. Consequently, regarding the content to communicate, each agent shall transmit its relative location and private reference $\mathcal{M}_t^i$ (i.e. map being explored up to now), with information of grids being visited and sensed (i.e. corresponding state values). Furthermore, as indicated in [6], state-value function with reward map shall be also sent. At the receiver end(s), after obtaining external private reference ($\mathcal{M}_t^j, j \neq i$) and experience, the robot will update the original private reference M_i, proceeding in two stages:

1. For each grid v in $\mathcal{M}_t^j$, use v_i to represent v according to its coordinate system. If v_i is not in $\mathcal{M}_t^i$, add v_i into $\mathcal{M}_t^i$. The corresponding action-value from agent u_j, $Q_j(v_j, a)$, will directly substitute u_i's action-value

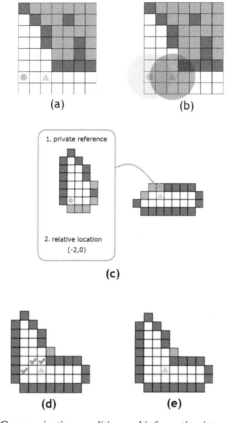

Figure 10.18 Communication condition and information integration procedure.

$Q_i(v_i, a), \forall a \in \mathcal{A}(v_i)$. Due to collaboration, agent u_i trusts u_j's experience.

2. For value of each grid on u_i's reward map $R^i(g)$, update reward map by comparing to u_j's reward map $R^j(g)$ and update with the smaller value because it indicates nonoptimal reward at that grid.

$$R^i(g) = \min(R^i(g), R^j(g)). \qquad (10.26)$$

Figure 10.18 describes a two-agent case when agents can communicate, and how they facilitate information assimilation. Subfigure (a) is the perception of environment. Let the triangle mark and circle mark represent agent u_i and u_j respectively. Subfigure (b) plots the communication range $r = 2$ of each agent. Since they are reachable from each other, they will start to exchange information. The left part of (c), including private reference of u_j (denoted by M_j) and its relative location, are the information sent to agent u_i. As shown in (d), u_i checks the external message with update rule. Grids with check mark obey rule (II), hence it will be modified as cleaned in M_i. Part of information in M_j is new for M_i to update by rule 2). Ultimately, agent u_i possesses a private reference in (e).

Again, as shown in Figure 10.20, wireless communication to exchange useful information between two robots indeed significantly improves the performance of collaborative MAS from every aspect, while limited difference in 100% completion time for two robots without wireless communication compared with single robot. Without communication, two collaborative robots pretty much work separately with repetitive efforts.

Random errors

In addition to ideal communication, a more realistic scenario is random error that cause packet drop. We assume random error happens at each communication link independently. Any transmission link, say from u_i to u_j, there is a probability e occurring random errors resulting from interference and noise. That is, a single directed transmission could fail with probability e. $\forall i, j \in \mathbb{N} : 1 \leq i, j \leq N, i \neq j$,

$$Pr \left\{ u_j \text{ receives from } u_i \middle| \overline{y_t^i, y_t^j} \leq r \right\} = 1 - e.$$

As expected in Figure 10.20, random errors deteriorate the completion time compared with ideal communication. However, the performance loss due to errors in communication is not as significant as the case of resource-sharing MAS in previous sub-section, which is not surprising since some failures to exchange information would not create significant loss for two collaborative agents in a rather large-scale and time-consuming mission.

Multiple access: p-persistent rt-ALOHA

For collaborative agents, multiple access communication in a form of mobile ad hoc networking is more realistic, while assuming a single communication channel being available. Multiple agents contend this multiple access channel resulting collisions to lose information exchange require multiple (or random) access protocol to coordinate transmissions. A fundamental difference for communication among robots or intelligent agents, compared with personal communication, traditional throughput-delay concept can not reflect the true need for communication, since each robot or agent must take an action based on real-time information in each time instant. The re-transmitted or backlogged information can be immediately obsolete. Therefore, a modification of slotted ALOHA, named as real-time ALOHA (rt-ALOHA) was recently proposed [6] (also in previous sub-section), which aligns the design trend of ultra reliable and low-latency communication (uRLLC) by grant-free access, no acknowledgement, and discarding re-transmission, that could finely support multi-agent learning tasks. The rt-ALOHA proceeds in the following procedure.

- When the channel is busy, the agent (i.e. cleaning robot) is ready to receive immediately.
- When the channel is idle, the agent broadcasts the message of desirable content, then immediately turns ready to receive from others right after transmission without any acknowledgement.
- There is no retransmission and thus backlogging.

From the study of random (packet) errors, persistent transmissions from $N \geq 3$ agents may create collisions and destroy the content of multi-access communication. To exchange useful private reference to effectively enhance overall performance of collaborative MAS, the p-persistent concept can be introduced to regulate rt-ALOHA, to create two operating modes to propose p-persistent rt-ALOHA borrowing the concept from CSMA :

- Proactive: if the agent senses other agents are within its communication range and the channel is not busy, agent broadcasts messages with probability p_p.
- Reactive: When the multi-access channel is busy, the agent stays at the reactive mode, ready to receive other's broadcast. When agent senses other agents are within its communication range, agent stays at the reactive mode with probability $1 - p_p$.

To present overall performance of collaborative MAS using p-persistent rt-ALOHA for multiple access communication, two scenarios of $p_p = 0.1$ and $p_p = 0.3$ are selected in Figure 10.19. By employing more collaborative robots with appropriate multiple access communication, the overall performance is satisfactory, say 10 collaborative robots using p-persistent rt-ALOHA saving around 80% time. Small p_p indicates less number of times to exchange information between agents, and hence leads to higher variance and unpredictable system performance, such as the case of $N = 5$ with $r = 5$ which can be worse than system performance of two agents. Actually, $p_p = 0.3$ shows pretty satisfactory performance for collaborative MRS. The

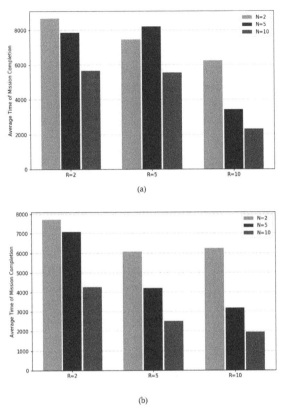

Figure 10.19 Performance of p-persistent rt-ALOHA (a) with $p_p = 0.1$ (b) with $p_p = 0.3$ for $N = 2$ (yellow), 5 (blue), 10 (red).

exact optimization or stabilization of the proposed p-persistent rt-ALOHA is left as future research in smart machine-to-machine (M2M) communication.

To have holistic understanding of the impacts from wireless communication to collaborative MRS, Figure 10.20 demonstrates the completion time from environment's aspect. All simulations are carried out with identical setting of two agents ($N = 2$) and communication range of two ($r = 2$) except for single agent case. p-persistent real-time ALOHA with $p_p = 0.3$, suitable for operating with less intense communication traffic, turns out having performance close to ideal communication. Exchange of proper information among collaborative agents using proper communication and networking methodology brings in overall system performance gain, which

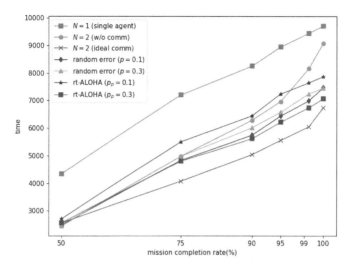

Figure 10.20 Average completion time for multi-agent systems, in terms of the percentage of mission completion, $N = 2$, $r = 2$ except the single agent case.

can be generalized to autonomous vehicles, man-robot collaboration, smart manufacturing, and robotics in general.

Figure 10.20 indicates a fact that collaborative MRS without networking performs approximately equivalent to a single agent no matter how many agents are operating simultaneously, while public/global reference is not available. This verifies that collaborative MRS not only benefit from parallel distributed computing, information exchange based on networking plays a key role to boost their performance, which has been overlooked for decades in artificial intelligence technology.

Networked MRS (or MAS) is still an active research area in robotics requiring further systematic investigations, and this section just provides an introduction with consideration from wireless communications. Actually, *distributed machine learning* may be a useful tool to study networked MRS, but it is not possible without significant expanding the scope of this book.

Further Reading: To explore further on multi-robot systems, please refer [3, 4]. Section 10.3 is based on [6, 7]. *IEEE Signal Processing Magazine* has a special issue related to distributed machine learning in May, 2020, for further reading.

References

[1] Stuart Russell, Peter Norvig, *Artificial Intelligence: A Modern Approach*, 3rd edition, Prentice-Hall, 2010.

[2] B.P. Gerkey, M.J. Mataric, "A formal analysis and taxonomy of task allocation in multi-robot systems", *Intl. J. of Robotics Research*, vol. 23, no. 9, pp. 939–954, September 2004.

[3] I. Mezei, V. Malbasa, I. Stojmenovic, "Robot to Robot", *IEEE Robotics and Automation Magazine*, pp. 63–69, December 2010.

[4] A. Koubaa, J.R. Martinez-de Dios (ed.), *Cooperative Robots and Sensor Networks*, Springer, 2015.

[5] K.-C. Chen, T. Zhang, R.D. Gitlin, G. Fettweis, "Ultra-Low Latency Mobile Networking", *IEEE Network Magazine*, vol. 33, no. 2, pp. 181–187, 2019.

[6] E. Ko, K.-C. Chen, "Wireless Communications Meets Artificial Intelligence: An Illustration by Autonomous Vehicles on Manhattan Streets", *IEEE Globecom*, Abu Dhabi, 2018.

[7] K.-C. Chen, H.-M. Hung, "Wireless Robotic Communication for Collaborative Multi-Agent Systems", *IEEE International Conference on Communications*, 2019.

Index